the GUINNESS guide to underwater life

Translated from the original French by
Michael Hopf

Technical assistance given by
Gerald L. Wood FZS,
Colin Doeg, British Society of Underwater Photographers.

Original copyright 1974 by
Editions Denoël, Paris, France.

© English Language edition 1975 by
Guinness Superlatives Limited

Published in Great Britain by
Guinness Superlatives Limited, 2 Cecil Court,
London Road, Enfield, Middlesex, England

ISBN 0 900424 58 3

Photoset in Great Britain by
REDWOOD BURN LIMITED
Trowbridge & Esher

Printed by A.I.P., Argenteuil, France

the GUINNESS guide to underwater life

Christian Petron and Jean-Bernard Lozet

guinness superlatives ltd.

All photographs shown in this book have been taken with the following films:
KODAK EKTACHROME X 64 ASA
KODAK EKTACHROME HS160 ASA
KODAK EKTACHROME EP 50 ASA
KODACHROME 11 25 ASA
KODACHROME X 64 ASA

Rorquals
in the sea off the Bermudas

Preface

For a long time, diving was the domain of a few specialists and enthusiasts. The conquest of space attracted more attention. Suddenly, there was an awakening: the continental shelf loomed up like an immense new territory on man's doorstep. Strangely garbed technologists began to invade the sea and, at the same time, swarms of people diving for sport probed ever deeper into the marine environment.

Slowly and inexorably, man is becoming master of his new domain. Today we are witnessing an explosion of ideas, and a proliferation of material dealing with the many aspects of this fabulous world. The scope of the subject is inexhaustible. One of the fundamental problems presented to the underwater stroller is how to identify the innumerable animal or plant species, often with strange shapes and wonderful colours, which meet his eyes.

This book, conceived by Christian Petron in collaboration with Jean-Bernard Lozet, is of particular interest and forms an indispensable guide for the diver. It enables him to put a name to specimens of flora and fauna, not only in the Mediterranean, but also in tropical waters, which are no longer reserved for the privileged few, but are visited each year by an increasing number of marine enthusiasts.

It is the dream of every diver to bring the lore of the sea into his home, and modern techniques for breeding fishes in aquariums produce remarkable results in this respect; this book, accordingly, takes this subject into account. It will prove very useful to all those who are eager to continue watching the fascinating life of the reef at home.

Finally, anyone who has ever relished the discovery of the 'sixth continent' feels the irresistable need to preserve on film this world in which he is a privileged onlooker. Underwater photography and filming are undergoing an unexpected expansion. In the third part of the book, Christian Petron, who is currently one of the foremost specialists in underwater photography, has made an extremely interesting and exhaustive study of this subject. This can be referred to by all those, ever increasing in number, who wish to record memories of their journeys at the bottom of the sea.

We therefore welcome the publication of this comprehensive and interesting work, with its original and attractive layout. I wish it every success.

Jacques DUMAS
President
of the World Confederation of
subaqua activities.

Summary

Pink coloured algae (*Asparagopsis armata*) shown
here in their sexual form. They appear in quantity
towards the end of spring.

These large greenish bulbs, which are often found on beaches after strong gales, are algae and not sponges, despite their shape.

These algae (*Acetabularia mediterranea*) grow in the shallows and reappear when the water becomes warmer.

15

Posidonia are very common in the Mediterranean.
These plants are usually clumped together forming
meadows under the sea.

2. Plant life

In the underwater world, the distinction between the animal and plant kingdoms is difficult to establish. This is the case with protozoans: some members of this group display all the characteristics of plants, whilst others, like animals, feed solely on living matter.

There is a large variety of underwater plants, which are either clearly visible to the naked eye (pluricellular organisms) or microscopic (unicellular). The latter are scattered throughout every ocean in the world, and constitute part of the drifting plankton which is at the mercy of the current.

In the sea, as on land, plants are capable of obtaining solar energy and of absorbing mineral elements, by photosynthesis, in order to transform them into organic matter which can be used directly by animals. That is why the limit of diffusion of light rays determines the threshold beyond which vegetable matter cannot survive. This limit is generally found at around 260–330 ft (80–100 m), and will only rarely reach 300–460 ft (100–150 m).

Amongst the pluricellular organisms, an important distinction must be made between algae and phanerogams.

Algae are typically aquatic plant organisms which have no need to fix roots in the soil to live. Some drift, whilst others become attached. Apart from the green pigment (chlorophyll), their cells have a large variety of pigments in differing proportions, which cause their colours to range from brown to red to blue. Others secrete a calcareous skeleton. They then spread out flat, which makes them look like corals of the same type.

Phanerogams are terrestial plant organisms which have migrated under the sea while still conserving their characteristics: roots deriving life from the soil, and reproduction by seeds and shoots.

Posidonia are a species of phanerogams, making up large water plant communities, which are very common in the Mediterranean. When temperature conditions are adequate, these plants become covered with little yellow flowers and little tufts of pink threads which, for a long time, were thought to be a separate species.

Asparagopsis multiply both by sexual means and through the intermediary of stem suckers with little barbed branches which break loose. First discovered in Australia in 1854, this species has apparently migrated very rapidly, and was prominent in France in 1925 and in Ireland in 1941.

The sp
palmat

Red co
to its sh
was hea

A gym
ous c

drawn to the very important role played by jellyfishes in the underwater biological life-cycle. In fact, they are used as a refuge by numerous species of fishes while they are growing into adults. It can even be stated that a young fish, a cod or mackerel, deprived of this protection, could not survive. Some Scyphozoan jellyfishes are edible, and are much sought after in Japan and some Pacific islands.

Corals Corals, gorgonians, sea-anemones and madrepores belong to the very large Anthozoan class.

The Alcyonacea belong to the Octocorallia sub-class (soft corals) i.e. they have eight tentacles. They usually form very colourful colonies, as species photographed in the Mediterranean bear witness: *Alcyonium palmatum* and *Parerythropodium coralloides*. One of them, the *Corallium rubrum* or Mediterranean red coral or jewelled coral, is particularly well-known and deserves a mention. It is indeed one of those rare coelanterates whose limey skeleton is completely pigmented. It is only to be found in the Mediterranean.

In ancient times and more especially in the Roman era, red coral was known and used as jewellery. Until recent years, it was fished either with nets or with special dredges, which were dragged along the rock faces. The yield was scanty, as the *Corallium rubrum* esconced itself in grottos, cracks, and other places sheltered from the light. The appearance of the heavy diving suit, followed by the aqualung, has given rise over the last thirty years to the expansion of intensive fishing, to such an extent that certain zones have been completely cleared. Fortunately, only those pieces of coral of a set diameter are saleable; thus allowing the smaller pieces to perpetuate the species. The price of red coral depends on the thickness of its branches and, once carved, can fetch several thousands of pounds. Then there is the *Corallium japonicum*, found in the Sea of Japan, which can weigh several pounds and measure more than a yard in breadth.

Gorgonians (Sea Fans) Contrary to their plant-like appearance, gorgonians are animal colonies which secrete a pliable and resilient horny substance called gorgonin, which forms the skeleton. This material, known as 'black coral', is also used in jewellery.

Gorgonians live at all depths, from the surface areas down to about 3250 ft (1000 m). Moreover, at these great depths, some species of gigantic dimensions are encountered. Once in full bloom, gorgonian polyps form a compact web which traps food particles borne by the current. Each polyp is equipped with eight pinnated tentacles. It is bound to the others by a common tissue: the coenenchyme, which is crossed by ducts and contains the characteristic calcareous spicules for each species.

The largest Mediterranean gorgonian (*Paramuricea clavata*) only appears in its glittering red mantle underwater when lit by an artificial light. When brought to the surface, it turns black. It is found on the great submerged slopes of the Mediterranean, and notably in the approaches to the islands and cliffs located between Cassis and Marseilles, at a depth of 100–200 ft (30–60 m). It is rarely seen at a depth of 50 ft (15 m), but when it is, it appears in the form of a spherical colony.

Every diver knows the gorgonian *Eumicella cavolini*, well. It is the most common of this species in the Mediterranean. It has a brilliant yellow colouring, which keeps for some time when exposed to air, and is found in shaded zones at a depth of from 10 to 130 ft (3 to 40 m). Like the *Paramuricea clavata*, it is fond of vertical rock faces lying exposed to the current, so enabling it to filter the greatest possible quantity of water.

Sea-Anemones The *Actiniaria*, more commonly known as sea-anemones, are hexacorallian, as they have six tentacles. Devoid of a skeleton, they sway about in the current looking like, and often mistaken for, pretty flowers. They seem to adapt very easily to the most difficult living conditions in cold water as well as at great depths, since one living specimen has been brought up from 29 000 ft (9000 m) down.

Then again, they have the characteristic ability to take on a certain number of commensal or parasitic animals. Among the latter should be mentioned the tropical shrimp (*Peridemenes brevicarpalis*), which lives in the heart of the sea-anemone, only leaving temporarily to clean the teeth or the wounds of certain fishes which come voluntarily to have themselves taken care of. Mention should also be made of the little clown-fish, which at the least alert, runs to shelter amongst the tentacles of the sea-anemone. Some recent experiments have tried to prove that the clown-fish *Amphiprion* knows how to 'tame' the sea-anemone by covering itself bit by bit with a protective mucus.

Another well known example of commensal association is found in European seas—that of the sea-anemone *Adamsia palliati* and Prideaux's hermit crab (*Pagurus prideauxi*). The latter, which has a soft, asymmetrical abdomen, lives in abandoned sea-snail shells and carries the anemone around on its shell. As it feeds, the anemone gathers fragments left by the crab with its thread-like tentacles. The anemone is armed with stinging cells and these protect the hermit crab from enemies like the octopus. When the crustacean outgrows its home it usually leaves the anemone behind. Sometimes, however, the basal disc of the anemone completely encloses the shell and, as the crab grows, so does the anemone, adding to the effective volume of the habitation. Thus, the shell does not have to be replaced.

Among the Mediterranean sea-anemones, there are some which spend their life hidden in the sand, only showing their tentacles, whilst others, like the *Anemonae aurantia*, stay naturally fixed to the rocks. This last-named species is especially sought for its delicate flesh.

The Ceriantharia are large sized sea-

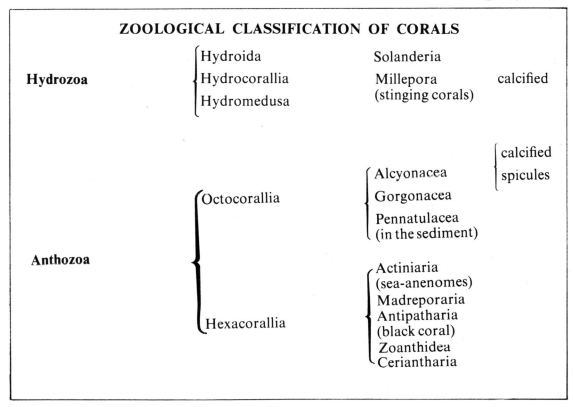

ZOOLOGICAL CLASSIFICATION OF CORALS

Hydrozoa
- Hydroida — Solanderia
- Hydrocorallia — Millepora (stinging corals) — calcified
- Hydromedusa

Anthozoa
- Octocorallia
  - Alcyonacea — calcified spicules
  - Gorgonacea
  - Pennatulacea (in the sediment)
- Hexacorallia
  - Actiniaria (sea-anenomes)
  - Madreporaria
  - Antipatharia (black coral)
  - Zoanthidea
  - Ceriantharia

Close-up of the tentacles of a sea-anenome (*Anemae aurantiaca*). Small shrimps, impervious to its stings frequently swim through the anenome

The sea rose (*Stertella septentrionalis*) is sought by divers for its decorative effects.

False coral (*Adeonella calvati* or *Chilostome ascophora*) is often found clustered into little bushes.

Tritons or marine conches (*Charonia nodiferum*) eat starfishes.

A Hexacorallian (*Cerianthus membranaceus*). It live: in the mud of harbours where it feeds on wast products accumulated in the polluted waters

Coelanterates (*Parazoanthus axinellae*) with some
polyps extended and others retracted.

anemones which secrete a comparatively rigid tube encrusted with particles, grains of sand and debris, and possess large pitted retractable tentacles which undulate in the current. The body is itself a tentacle, with numerous stinging capsules. Some species, common in the Mediterranean, can attain impressive dimensions.

Madrepores (True Corals). They are very widely distributed in tropical seas, and are much rarer and more scattered in cold and temperate seas, where they never form very extensive colonies. In the Mediterranean two types are relatively common: firstly, the *Cladocora caespitosa*, which is found particularly in certain deep creeks of Cassis and which forms a small independent colony, sometimes reaching 20 in (50 cm) in diameter: secondly, the *Coenocyathus*, which lives scattered around rock faces at a depth of 30–100 ft (10–30 m), together with gorgonians and red corals.

False Corals Bryozoans or (false corals) are often mistaken for real corals. These microscopic animals construct little shells, mostly calcified, placed side by side, which communicate with each other through cavities. In this way, food ingested by one of them benefits the whole colony. Bryozoans are also characterised by their ability to attach themselves with ease to all kinds of substrata.

The sea-rose (*Stertella septentrionalis*), with its delicate and finely woven lacework, likes shaded rocks and often attaches itself to gorgonians.

The 'false coral', scientifically termed *Adeonella calviti* and *Myriapora truncata*, frequents the coral-bearing reefs of the Mediterranean.

There are cases of epibiosis, i.e. attachment to another animal. Some colonies can attach themselves to the tails of certain Pacific sea-snakes. False coral also likes to hook itself onto the hulls of boats.

Among the coelanterates yet to be mentioned are the Zooxanthellae, which make up a sub-group of the Hexacorallia, having no calcareous skeleton but never-

theless forming part of the colonies. The *Parazoanthus axinellae* is a very fine species which often lines shaded rock walls and Mediterranean grottos, glowing brilliant yellow in the slightest light.

Molluscs

Of all the wonderful shellfishes in the world, those found in the Mediterranean must remain poor relations, for whatever their attractions in shape and colouring, there is always a larger and more vivid distant cousin in the Tropics to outshine them, which explains why collectors have not shown much interest towards them. However, some enthusiasts obtain outstanding results in this field. For instance, Mme Kety Nicolay, director of the Roman conchological review 'La Conchiglia', has been able to compile one of the most remarkable collections in the world.

Sea mussels (Murexes)

The history of the shellfish, from ancient times, has been bound up with that of the Latin people. Industries were actually built up in places where certain Mediterranean mussels were found. A mucus was extracted from them, which formed the basis for a purple dye sold throughout the western world and at Tyre it is still possible to see the great cauldrons in which they used to soak the fabrics for dyeing.

Seapens

From seapens, a valuable material is also extracted. In fact, the majority of bivalves are anchored to the ground by very pliable and resilient little filaments, the byssus threads, which, when suitably treated and woven, have been used for centuries to make the most luxurious garments.

Conches

Tritons or conches have been used by man since time immemorial for such things as a loud siren used by fishermen

28

and hunters, and as a religious symbol (paintings and drawings of it are found in many ancient holy places). It was also used, more prosaically, as a kitchen utensil, as a container, and of course, like ninety per cent of shellfishes, as a foodstuff which is still considered a delicacy nowadays.

Tritons are very common all over the Mediterranean, with the possible exception of colder waters in the northern-most part, the French and Italian Coasts. On the other hand, they are often fished in the Balearic Isles, the south east of Spain, and in North Africa: they get caught up in the nets when they come to 'pillage' the captive fish. Conches largely feed on sea-stars, just like their tropical cousins.

Octopuses

Cephalopods consist of three sub-orders: the Octopoda (octopuses) with eight arms, the Decapoda (cuttle-fishes and squids) with ten arms, and the living fossil Vampyromorpha, which also has ten arms.

The common octopus (*Octopus vulgaris*) is imbued with rare intelligence. Experiments have shown that it is capable of memory and reason in terms of a clearly defined idea.

It leads a quite extraordinary life. The spawning periods are the opportunity for long amorous disputes, which end with the male surrendering to the female one of his eight arms, the hectocotylus, which contains the sperm packet. She then places it in the shelter of her mantle cavity, in which the eggs to be fertilised are located. Having found herself a hole, she sticks her eggs into bunches and watches over them without eating whilst airing them with her syphon until they hatch. Then the exhausted mother dies.

There are several species of octopuses in the Mediterranean. Besides the *Octopus vulgaris*, there is the *Octopus macropus*, differentiated by its smaller size and its more slender tentacles; the *Octopus de filippi*, very small with long

Eggs of an octopus.

Octopuses swim by jet propulsion: violently expelling water from the sac by means of a syphon.

29

An octopus (*Octopus vulgaris*): a friendly little animal
which is often depicted as a monster by authors such
as Victor Hugo.

A common cuttlefish (*Sepia officinalis*) which like all
cephalopods has a parrot-shaped beak.

The effective camouflage of a
cuttlefish displays its remarkable
gift of mimetism.

Eggs of a cuttlefish. The baby
cuttlefishes can be seen in their
capsules.

tentacles, and the Eledone, (lesser octopus) which can be recognised by the single row of suckers on its tentacles, whereas other octopuses have two. Two species of eledones are encountered: the commonest is *Eledone cirrhosa*. The *Eledone moschata* or muscat octopus, is distinguished by its white dots and strong smell of musk.

Cuttlefishes and squids

This family of molluscs possess ten arms, two of which (tentacles) are hidden by the other arms around the mouth and can be catapulted onto their prey.

Cuttlefishes are endowed with an unusual power of mimesis and often settle in the sand like flat fishes. At the least alarm, they emit a particularly dense cloud of black ink in which they lie hidden.

During the spawning period, the male of the species behaves with extraordinary tenderness towards the female, caressing her with his oral tentacles. The eggs are laid in little white or black pouches, which are assembled in clusters and left under a stone. The little cuttlefishes, only a few millimetres in size, are not long in hatching and are already able to emit their ink at the least danger. Unfortunately, many of them fall prey to wrasses or other small predators.

There are three main species of cuttlefishes: *Sepia officinalis*, *S. orbignyana* and *S. elegans*.

Squids live in compact shoals, and at night they come up to the surface looking for shrimps. They are frequently encountered in the Mediterranean, where they are intensively fished. Phosphorescent particles on their skin attract the prey, just as lures would do.

Crustaceans

The class 'Crustacea' belong to the phylum Arthropoda, which contains more species than the rest of the animal kingdom put together.

The word 'crustaceans' brings to mind crawfishes, lobsters, crayfishes or crabs. In fact, this vast class, comprising some 25 000 members, embraces the Krill (immense shoals of pelagic crustaceans which are the staple diet of the baleen whales) as well as the hundreds of species of little mites, the numerous parasites of aquatic animals, and a large proportion of plankton.

Although the majority of crustaceans are aquatic animals, some live in fresh water and others have even completely adapted to life on land. However, it is only the decapod crustaceans which are well known. Like crawfishes, they have five pairs of thoracic limbs. Decapods are further divided into two sub-orders, depending on whether they swim (Natantia e.g. the shrimp), or crawl (Reptantia e.g. the crab).

The Natantia are basically known and represented by shrimps, which are remarkable aquatic animals. They constitute an extensive reserve of food for man, who at times catches them in tens of thousands of tons (e.g. *Peneus setiferus* in Mexico). They also have rather unexpected functions in the biological cycle. They act as dentist to the fishes who come to have their teeth cleaned and their wounds cleansed. Fishes have actually been seen queuing in front of the den of these astonishing shrimps which, thanks to their usefulness, are spared by the majority of fishes.

The Reptantia are divided into three main branches, according to whether their tails are long, short, or limey. Among the Reptantia with long tails, the most famous are the lobsters, crawfishes, crayfishes and squill-fishes. The common or European lobster (*Homarus vulgaris*), which can reach a weight of 20lb (9 kg), is the uncontested king of crustaceans owing to its gastronomic properties.

Crawfishes and squill-fishes (*Patinurus* and *Scyllarus*) are distinguished from lobsters by their absence of pincers. While crawfishes are fairly common,

The European lobster (*Homarus vulgaris*), which has pincers, should not be confused with the crawfish, which has none.

A Galathea (*Galatea strigosa*), noted for the nimble way in which it nips in and out of the rocks. Its bitter taste can ruin a good fish soup.

A Mediterranean crawfish (*Palinurus vulgaris*) often found in the springtime along the rocky slopes where it comes to settle.
Its tasty flesh makes it the most sought after of all crustaceans.

squill-fishes on the other hand are much less so, but are much sought after and appreciated by gourmets.

Hermit crabs (*Anomura*) are often confused with gasteropods. They are in fact always found in a shell, which they use as a shelter. They are not builders, just temporary tenants, until their rate of growth compels them to look for a larger shell.

Crabs (*Brachyura*) are legion throughout the world, in cold as well as warm waters.

The Mediterranean harbours a great quantity of very interesting species. Two of the most extraordinary will now be mentioned.

One of these crabs has flattened claws which interlock perfectly on the front part of the animal. It then looks very much like a coxcomb (hence its nickname 'crested crab'), and is protected in this way by an ideal shield. It generally lives in the sand, in which it can quickly bury itself.

As for the *Dromia vulgaris*, this has a very peculiar deformation, namely the backward displacement of its four rear limbs, two of which are practically on its back, sticking up in the air. These limbs are used to support permanently a sponge, which camouflages it and ends by fusing perfectly into the shape of its body. If, in the course of flight, it is forced to desert the sponge, the 'crab' will always return to look for it.

A little squill-fish (*Scyllarus arctus*) good for flavouring Provençal fish soups.

The spider-crab lays its eggs in the shallows during the month of May.

The hermit crab (*Pagure Pagure*) lives in symbiosis with a sea anenome which attaches itself to its shell.

Red starfish (*Echinaster sepositus*), the most common
starfish in the Mediterranean.

Red velvet swimming crab (*Portunus corrugatus*). It
swims by using its webbed rear claws.

Echinoderms

Present in every sea in the world, echinoderms are a very complex phylum, and despite their efforts, zoologists have not yet been able to clearly classify them into the various branches of the animal kingdom. However, it seems likely that their primitive shape makes them close cousins of stomatopods (mantis shrimps). Furthermore, these animals have been divided into five classes: Starfishes (Asteroidea), brittle stars (Ophiuroidea), sea urchins (Echinoidea), sea cucumbers (Holothuroidea), and sea lilies (Crinoidea).

Starfishes

The Asteroidea or starfishes, have become famous through a number of films showing them in the process of attacking innocent scallops, which flee in panic.

The majority of them (in particular *Haulia echinaster* and *Marthasterias*), turn their stomach inside out, which enables them to cover and then eat their prey: sponges, scallops etc. Others (*Astropecten*) directly ingest their victims, usually little shellfishes whose shells are then discarded. It is not unusual to see its imprints surrounded by empty shells.

Sea urchins

The *Spatangus purpureus* is a heart urchin which basically lives entrenched in the sand. Although it is easy to find its empty shell, it is difficult to uncover the odd spines which sometimes slightly stick out from the sandy beds at a depth of 16–200 ft (5–60 m).

The *Stylocidaris affinis* is one of those rare species of Pencil sea urchins, which are more common and larger in tropical seas.

The *Sphaerechinus granularis*, a more familiar species, is characterised by its white spines. It is most frequently encountered at depths of between 65 and 165 ft (20 and 50 m).

Sea cucumbers

Not very attractive and often called rude names, sea cucumbers are still worthy of interest.

Their cylindrical body is made from a soft tegument strengthened by a loose calcareous skeleton, terminating at one end with the mouth, which is surrounded by tentacles, and at the other with the anus.

The sea cucumber has a complex defence system. It involves cuvierian organs (kinds of fibres, filled with toxin, emitted through the anus), which break loose once they have struck their prey. Subsequently, these organs are quickly regenerated. Some sea cucumbers, which do not have these organs, directly diffuse the poison through their bodies. The toxin emitted by sea cucumbers is relatively potent. In a few minutes it can kill every creature in an aquarium, and can even blind a man if splashed in his eyes. Its repulsive action has even been shown to work efficiently on sharks.

On the other hand, sea cucumbers are quite edible, and are much sought after in some countries, especially in the Orient, where they are known as 'trepang'. Moreover, the membrane surrounding the gastric tract is a real delicacy.

Chordates

The primary group of chordates includes vertebrate animals and those species possessing only one dorsal rod (the notochord), a flexible axis which ensures that the body remains relatively rigid.

Sea squirts (*Ascidiacea*) are invertebrates belonging to this group. They inhabit the oceans, either on the surface or at great depths. They subsist by filtering water, which is passed through the gill slits of the pharynx into the atrial cavity before being expelled with the waste products through another orifice. They can live either freely or attached. The larvae drift until they reach maturity, at which time they secrete a semi-rigid tunic, which

Top left: A sand sea urchin (*Spatangus purpureus*): its shell can only be seen on sandy beds once it is dead. Whilst alive, it buries itself in the sand to look for micro-organisms on which it feeds.

Top right; This starfish (*Astropecten aurantiacus*) lives half embedded in the sand beds.

Middle left: Pencil urchin (*Skilocidaris affinis*) living at a minimum depth of 165 ft (50 metres).

Middle right: An echinoderm: detail of the cuvierian organs of a sea cucumber.

Opposite: An echinoderm (*Holothuria Forskali*) commonly known as a sea cucumber.

41

A colony of sea squirts (*Eudistoma sp.*) closely resembling a sponge. It is however differentiated by its more rigid texture.

A colony of sea squirts (*Diplosoma gelatisiosa*) attached to a gorgonian.

A colony of little sea squirts (*Clavelina lepadiformis*), their remarkable graphic structure is brought into relief by a fine white line.

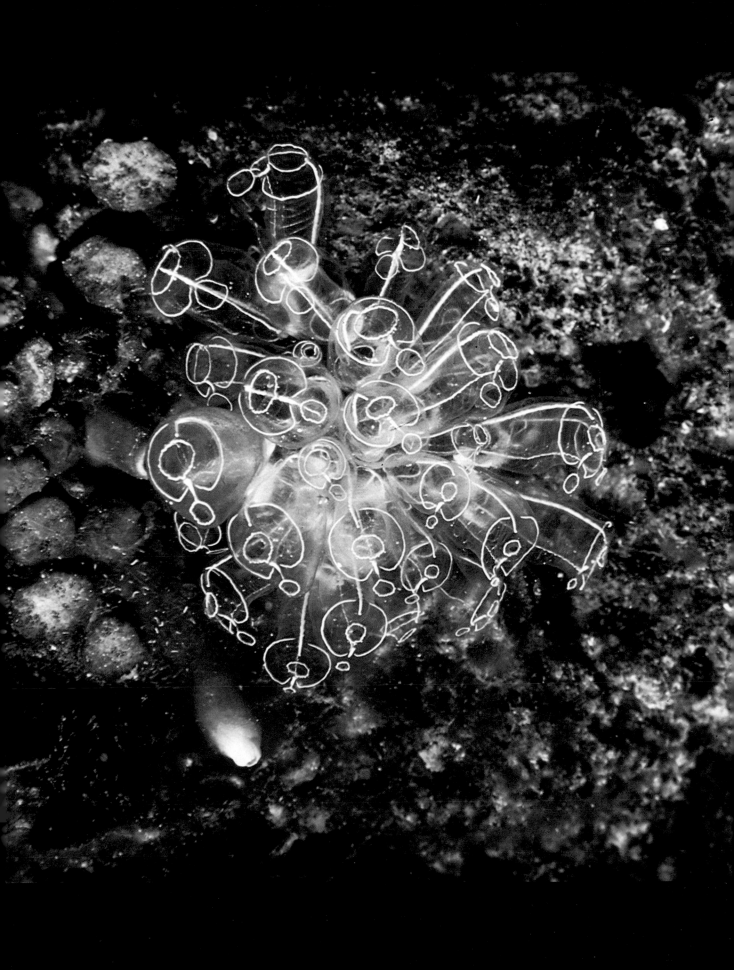

has earned them the name 'tunicates'. This characteristic is particularly noticeable in the *Clavelina lepadiformis*.

Sea squirts reproduce sexually, either by giving birth to larvae or by emitting stolons (creeping suckers) on which individuals separated from their parents can bud. They can form themselves into colonies by fusing their tunics together.

Dogfishes

The dogfishes or cat sharks are the most common types of small shark found in the Mediterranean. The lesser-spotted dogfish (*Scyliorhinus caniculus*), whose greatest length is 32 in (80 cm) lives mainly along the coasts of Sardinia, Italy and France. The greater-spotted dogfish (*Scyliorhinus stellaris*), which is more than 3 ft (1 m) in length, prefers to move about in the rocky deeps.

These fishes are nocturnal predators, thoroughly adapted to life in the shades. They are equipped with nictitating eyelids (only opening in the dark), which makes them blind in the daytime. During the day they are sometimes found deeply embedded in a crevice, with their tail outside, somewhat reminiscent of the nurse and sand sharks of tropical waters. If anyone cautiously pulls their tail, they clumsily swim away, knocking blindly against every rock in their path. Dogfishes, like all sharks, are dangerous predators. Professional fishermen do not much care for them, as they can cause substantial damage to their ground nets when they come and tear at the fishes with their sharp little teeth.

Dogfishes are an oviparous species. At spawning time, they mate on a rock, the male huddled against the female while surrounding her with his tail. The embryonic development lasts for 9 months. The eggs are laid in twos, in lozenge-shaped capsules, which cling by means of long threads to gorgonians, drifting with the currents. Little by little, the embryo develops. The capsules are covered with aquatic concretions which protect the egg from predators.

One night the baby dogfish will emerge alone from its capsule, often with its yolk sac not reabsorbed. It will immediately hide in the shelter of a crevice, while it waits to be completely safe from attack before facing a difficult life in which the young hunter is often itself hunted.

44

Dogfishes are not dangerous: they can be grasped by the tail and held above the head so as to avoid being bitten.

Eggs of a dogfish on the day they were hatched. The threads which enable the eggs to cling to algae and gorgonians are visible.

In the second photograph, the capsules can be seen becoming naturally covered with concretions.

The dogfish is hatched before it is ready: still in its vitelline sac, it hides in a crevice waiting until this nutritive pouch is completely reabsorbed.

At the end of the cycle, the young dogfish is ready to face life in the role of a nocturnal predator.

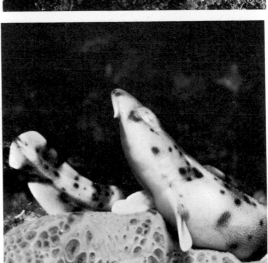

Eyelid of a dogfish: the iris closes automatically on
contact with light.

46

The blue sting ray (*Dasyatis violacea*) gives the diver
the impression that it is flying through the water.

Sardines (*Sardina pilchardus*) often form very dense shoals.

Thornback rays (*Raja clavata*) are sought for their flesh.

Rays

Fearsomely equipped by nature, rays give divers the impression that they are flying in their aquatic surroundings, by making their flexible pectoral fins undulate. Although usually of moderate size near coastal stretches, their weight can reach several hundreds of pounds on large sand banks in the open sea.

Rays feed on small crustaceans and shell-fishes which they grind with their little rasp-like teeth. They all have a liking for sand, in which they can hide very effectively, only letting their eyes peek out. The two most widely scattered species are the thornback ray (*Raja clavata*) and the blue stingray (*Dasyatis violacea*). The latter has a tail armed with venomous saw-edged spikes.

Some of the females lay eggs in the shape of little pouches fitted with twisted fibres, made from a horny substance, which enables them to cling to algae and gorgonians. On hatching, the young fishes are already armed with their dangerous spines. Others are viviparous and simultaneously give birth to twenty perfectly formed alevins.

There are other kinds of rays: the famous electric ray (*Torpedo*), which has a veritable electrical system capable of emitting a charge of several hundred volts: the common eagle ray (*Myliobatis aquila*): the Bordered ray (*Raja alba*): and, rarer still, the starry ray (*Raja asterias*). All these animals are to some extent armed, and it is advisable to take the maximum precaution when approaching them.

Sardines

The commonest type *Sardina pilchardus*, follows the nocturnal migrations of plankton and comes up to the surface to feed on them. That is why it is only fished at night, with nets, lamparas, and even, despite restrictions, with dynamite. This last method causes almost irreparable damage, as can be seen on the Sardinian and Italian coasts, where animal life has been almost completely destroyed as a re-

sult of this abominable procedure. In Marseilles, despite official prohibition, the authorities, although quite aware that dynamiting is carried out by professionals, have not yet made any serious attempt to stop them.

When diving, it is unusual to sight a shoal of sardines during the daytime, but this can occur in open sea at the end of the afternoon, in plankton-filled waters.

Other less well-known species of sardine include the sprat (*Clupea spratus*) and the allis shad (*Alosa alosa*).

Rocklings

Rocklings, 'motellas' or forkbeards (*Phycis phycis*) are members of the small family Gadidae, belonging to the order Anacanthini. They are sedentary fishes. Having chosen their den, they remain there for several years. Rocklings are equipped with particularly prominent tactile organs on which they rest, like the celebrated tripod-fishes of the ocean deeps. Rocklings are nocturnal predators.

Moray eels

The common moray (*Muraena helena*) is the only species of Moray eel living along the Mediterranean shores. It can reach 5 ft (1·50 m) in length and weigh about 25 lb (12 kg). Like its tropical cousins, this eel is very powerful and is armed with retractile teeth. During the day, it stays lurking in rock cavities or in the breakwaters of harbour entrances; at night, it leaps onto over-confident fishes and devours them. It has a special liking for octopuses, with which it fights fiercely, and always emerges as victor with an entire arm in its mouth.

Contrary to what the Romans have taught us, the moray eel does not attack man. However, should a person happen to place his hand at the bottom of a hole while looking for a crawfish, and carelessly put it in the mouth of a moray, it will certainly bite him as a defensive reflex.

Moray eels do not survive easily in

The coloration of the grouper varies according to its surroundings.

Close-up of a grouper.

aquariums, but if they are kept there, they should be fed a live octopus every month —which will be adequate. These animals are extremely susceptible to diseases and parasites. When free, they always live in their hole in association with a group of cleaning shrimps of the *Stenopus* genus, which conscientiously pick their teeth and rid them of parasitic isopods. In an aquarium, deprived of this cleaning service, the teeth of the moray eel become infected and abcesses appear.

Conger eels

These large fishes are bluish in colour, and are very common in all temperate and tropical seas. They are particularly fond of the crevices of rocky coasts.

Conger eels can live at all depths, from the surface down to several thousands of feet where Dr Piccard's bathysphere has photographed them. Their size and weight can vary enormously from a few inches up to 10 ft (3 m), and from a few oz to 175 lb (80 kg).

The conger eel is a nocturnal predator, and stays in its hole during the day. It is fished with trawl lines or eel pots. Having a snake-like shape, it is exceptionally difficult to catch it in a net, which it severely damages when coming to eat the captured fishes.

These animals spawn in open sea, but it is unlikely that they spawn, like freshwater eels, in the Sargasso Sea. During this period, they undergo a true metamorphosis, enabling them to lay millions of eggs. They die of exhaustion a very short time after this phenomenal laying of eggs. The larva (Leptocephalus) leads a pelagic life (in deep water) for some time, and finally takes up residence near the shore where it adopts its true adult shape.

Groupers

The grouper (*Epinephelus*) is the most familiar of diving fishes and is not by nature ferocious. If left to itself it can even be tamed. It usually lives in rocks. Being non-migrant it finds a hole, where it can stay for several years if undisturbed.

The Mediterranean species, *E. guaza*, can live as easily at a depth of several hundred feet as it can in a few feet of water near algae. That is why it is a choice quarry for underwater hunters. This intensive pursuit has made the grouper almost disappear from the shallow areas of our shores, and it now usually remains at a depth of 130–230 ft (40–70 m). Being largely a predator, the grouper likes crawfishes and octopuses. Secure in its power, it has no enemies except man. This intelligent fish never lets itself be caught in a net. When wounded or facing a particularly dangerous situation, it quickly submerges to the deepest part of its cave and jams itself firmly between the rocks by opening its gills and erecting its dorsal spines. Man should avoid hunting these likeable dwellers of the depths, because there is unfortunately only one underwater reserve, at Port-Cros, where hunting is forbidden, and where the grouper can come and breed in order to lead a peaceful life in the shallows.

There are other less common species of groupers that are particularly found along the eastern shores of the Mediterranean. The grouper *Epinephelus alexandrinus* is especially met with on African coasts. The white grouper (*Epinephelus aeneus*) and the black grouper (*Epinephelus caninus*) are more rare, as is the red hind (*Epinephelus rubra*), chiefly encountered along the Turkish and Libyan coasts.

Sea basses

The sea basses *Serranus cabrilla* and *Serranus scriba* are well known to fishermen using trawl lines for their quickness in leaping on baits. They never become very big and do not exceed 12 in (30 cm) in length.

Being sedentary fishes, they take to themselves a territory which they do not leave until spawning time. They can be easily observed under water because their curiosity, doubtless triggered off by greed, is such that they actually follow divers.

52

Moray eels (*Muraena helena*) have retractile fangs, some of which are fed by poisonous glands.

Conger eels (*Conger conger*) are large nocturnal predators. During the day, they lie hidden in their grottos, with only their snouts peering out.

When they are alive, red mullets (*Mullus surmuletus*) are not red: in fact, they only assume this colour once they are dead.

Damsel fishes (*Chromis chromis*) often form veritable clouds along the rocky slopes.

Barbers (*Anthias anthias*) are characterised by a deeply forked caudal fin. They are distant cousins of the Serranidae family.

Porgies, saupes and picarels

Porgies abound in the Mediterranean. They are small, oval-shaped fishes, with short stout heads. Their jaws contain two rows of incisors.

Porgies are hermaphrodites: when dissected, they have been found to possess simultaneously both male and female gonads, so enabling them to change sex. They live communally, often around posidonian plant communities. They always stay near a hole in order to shelter all together at the least danger: in this way, they become a choice quarry for underwater hunters.

Two species live in shoals on the rocky Mediterranean coasts: the two-banded porgy (*Diplodus vulgaris*), decorated with a transverse stripe on its tail and longitudinal bands on its flanks; the white porgy (*Diplodus sargus*), which has vertical bands over its entire body.

Saupes or bogues belong to the same family as porgies—the Sparidae. They are very common all over the Mediterranean. Being herbivorous, they move around in compact shoals in order to browse on the algae which cling to the rocks. These fishes are streaked with yellow and green horizontal bands. They are very timid and difficult to observe at close hand as they flee at the least danger. Their flesh is not very edible, yet is appreciated by those people who like its pronounced taste of algae.

The bogue or 'poor man's sardine' (*Boops boops*) lives in shoals like the sparids along the rocky submerged parts of cliffs or in the open sea. Being omnivorous, they feed on mussels as well as on plankton. Fishermen are not too keen on them, as it is difficult for them to catch anything else other than these fishes if a shoal of them appear in the fishing area.

The blotched picarel (*Manea manea*) bears a resemblance to the Sparidae family in its general appearance, but differs from it in its way of life. It follows planktonic migrations and lives on the sea bed during the day. It has a protractile mouth, enabling it to trap in flight the little planktonic copepods, of which it is very fond. Being a hermaphrodite, it first passes through the female stage. At spawning time, it burrows little depressions in the mud, in which it lays its eggs. The male then fertilises and guards them until they are hatched. The spawning period falls between May and September, depending on the locality.

Red mullets

There are two types of red mullet: surmulets living on rocks (*Mullus surmuletus*), and the unstriped red mullet (*Mullus barbattus*), living in the mud.

These little fishes are distinguished by their tactile and olfactory barbels with which they search in the mud for little worms and crustaceans on which they feed. Unlike the tropical species (*Pseudupeneus*), which they closely resemble in shape, Mediterranean mullets have no teeth.

Spawning takes place in the spring; the young hatch in open sea and are a turquoise blue colour like sardines. They only assume their adult colouration, green and yellow with a red stripe, when they reach the shore line. After death, they turn red.

Grunts, drums, croakers

Distant relations of the sea basses, the Sciaenidae constitute a family whose different species are often difficult to identify. Their characteristics are as follows:

An elongation of the soft part of the dorsal fin.
Teeth only on the lower jaw; the upper palate has none.
They emit sounds by means of abdominal muscles attached to the air bladder.

Full identification of species of the Mediterranean Sciaenidae has not been achieved. However, it is possible to identify the corb by its double dorsal fin, its brown back, and its yellow belly; the umbrine by a short barbel under the jaw

56

and by its squat shape; and lastly, the maigre by its large size.

The corb (*Corvina negra*) always lives near rock faults or recesses. Hunters know them well, as they enable them to detect the hideouts of the grouper. Some people collect otoliths or ear-stones —little bones situated above the temples of the corb.

Cardinal fishes

The cardinal fish, *Apogon imberbis*, is the only representative of the Apogonidae family in the Mediterranean. This little animal always lives in the dark recesses of grottos and caves, sometimes in pairs, sometimes singly.

At spawning time, a veritable display takes place between male and female, which ends by the eggs being laid in the mouth of the male (a little mass containing nearly 20 000 eggs). The male guards them in this way while oxygenating them until they are hatched. These fishes are 'buccal incubators'. Nevertheless, unlike their tropical brothers, the young fishes do not seek shelter in their 'father's' mouth when faced with danger.

If this little fish is kept in an aquarium, it is advisable to operate different stages of decompression on it, as it is extremely fragile when brought to the surface.

Barbers *(Anthias)*

The *Anthias anthias* is a rare Mediterranean member of the family of sea basses, a very important species in tropical seas.

Barbers live deep down and are often fished with a trawl line by those with a taste for fish soup. They travel about in large groups over very precisely defined areas, which are located at a depth of between 100 and 165 ft (30 and 50 m).

Each female lays her eggs in a nest, while the male exercises a vigilant watch all around until they hatch. Should an enemy or predator approach the nest, all the barbers form a united front against the adversary.

In an aquarium, this splendid little pale pink fish feeds on shrimps.

Corb (*Corvina nigra*). Corbs and groupers are usually found together. The corb is fast disappearing from Mediterranean shores.

Deep sea barbers (*Callanthias ruber*) live on rocky beds.

57

Saupes (*Boops salpa*) travel about in shoals, grazing
on algae on the rocks.

The two-banded porgy (*Diplodus vulgaris*) is a choice
prey for underwater hunters.

There is another species of 'anthias' often met with in rocky depths: the *Callanthias ruber*. It is characterised by a deeply indented caudal fin, and never comes up more than 100 ft (30 m) from the bottom.

Damsel fishes

Damsel-fishes (*Chromis chromis*), belong to the Pomacentridae family, and create veritable clouds along the rocky coasts. They feed exclusively on little planktonic animals, and it is difficult to give them proper nourishment in an aquarium.

At spawning time, like the Maenidae, the females come and lay their eggs in nests built by the male, who immediately fertilises and watches over them until they are hatched. The turquoise-blue colour of the young fish gradually disappears as they grow older.

Blennies and gobies

Blennies (*Blennius gathorugine*) and gobies (*Gobius minutus*) are usually found in shallow waters. They are encountered in every port in pools left by the low tide, as well as in the still waters of river mouths.

They spend almost all their lives on the sea beds. They lay claim to a territory and travel through it in successive bounds, with long pauses in between.

Gobies are extremely voracious little animals, always looking for food. One need only wave a little worm in front of them, and they rush to it. They are not at all timid when in a pool of water, and if one has enough patience, they will come and eat out of the palm of the hand.

Their mode of life, severely threatened by pollution, no longer allows them to spawn in favourable conditions. Their flesh is insipid; and apart from being amusing to children, they do not attract the attention of fishermen.

Scorpion fishes

There are three main species of scorpion-fishes in the Mediterranean: the little brown scorpionfish (*Scorpaena porcus*), the chestnut or red scorpionfish (*Scorpaena scofra*) and the crested scorpionfish (*Scorpaena ustulata*).

They are endowed with amazing powers of mimicry. Not only are they able to assume the colour of their surroundings, but they can even camouflage their heads so that they merge with the rocks, where they lie in wait for little fishes, leaping out on them, as quick as a flash, with mouths wide open.

The chestnut scorpionfish can become very big and can weigh up to 18 lb (8 kg). It prefers the ocean depths, and more especially waters 100–130 ft (30–40 m) deep.

Brown scorpionfish are encountered along the breakwaters of harbour entrances, generally in the dark hollows at the intersections of the boulders.

When they are being fished, they must be handled very carefully. Their spines are in fact armed with a poison-filled gland which can induce extremely painful symptoms of blood poisoning. The chestnut scorpionfish are fished with a ground net, but they can also be caught with a trawl-line. It is more productive to fish for them at night, when they move around to hunt.

Wrasses

Distinguished by an elongated body, a retractile mouth with thick lips, and a long single dorsal fin, wrasses are represented in the Mediterranean by three species: the rainbow wrasse, the cleaner or crenilabruse, and the large wrasse.

Originally, the female rainbow wrasse (*Coris julis*) was thought to be a different species called *Coris giofredi*. Its hermaphrodite stage was noticed in an aquarium, when on changing sex it completely changed its skin.

The Turkish wrasse (*Thalossoma pavo*), another species of rainbow wrasse, undergoes a change of skin (involving a disappearance of its stripes) on passing into the adult stage. There is no difference between male and female adults. They are often encountered along the eastern

The size of the striped blenny never exceeds 2¾ in. (7 cm). It can be identified by the black stripe along its back.

The blenny (*Tripterygium tripteronotus*) has two little support fins to help it to alight on rocks.

The *Gobius minutus* is completely yellow and found at the foot of underwater slopes.

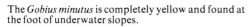

The *Gobius bucchichii*: king of the harbour beds.

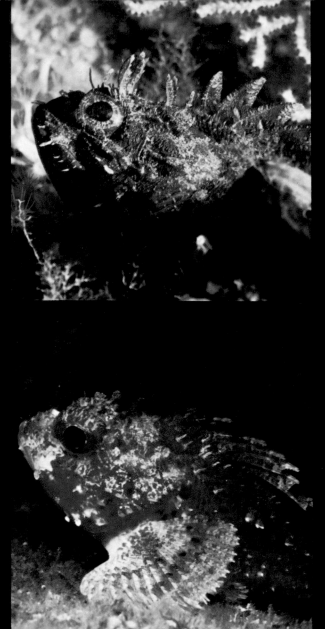

▲ A rainbow wrasse (*Coris julis*) in the male stage, surrounded by wrasses in the female stage.

Top left: the scorpion-fish (*Scorpaena porcus*) has a dorsal spine fitted with a poisonous gland which can cause grievous injuries.

Middle: the chestnut scorpion-fish (*Scorpaena scrofa*) can weigh up to 18 lb. (8 kg).

Bottom left: crested scorpion-fish (*Scorpaena ustulata*) so called because of its little skull-cap, the colour of which contrasts with that of the rest of its body.

Cuckoo wrasse (*Labrus bimaculatus*) in the male
▼ stage.

Tompot blenny (*Blennius gattorugine*): one of the rare
Mediterranean fishes to possess binocular vision.

Mediterranean and North African shores.

Some species, like the cuckoo wrasse (*Labrus bimaculatus*), rid the sides of other animals of parasites (Copepoda and Isopoda). Their skin, like that of the rainbow wrasse, is subject to variations in colour.

The larger wrasse or green wrasse (*Labrus turdus*) is very common in sea grass (posidonian) meadows, where they are found covered with numerous little parasites. They frequently have themselves 'deloused' by their little neighbours, the cleaners. Sexually, their appearance varies little. They are fished with a groundnet, rarely biting at the line on account of their small snouts which are best suited for scraping off algae.

Tunnies

Being a migratory fish, the tunny is large and strong, spindle-shaped and sheathed in a thoracic shell of large scales.

The common tunny (*Thynnus thynnus*) can reach 13 ft (4 m) in length and weigh 1650 lb (750 kg). Its skin glints with fine steely blue and silvery white hues. It travels across the Atlantic, but comes to breed in the Mediterranean, on the Spanish coastlines.

The red tunny sets up its spawning grounds along the Sardinian, Sicilian and Tunisian coasts. The eggs are of the planktonic kind and are fitted with an oil sac which enables the embryo to develop.

The tunnies are indefatigable swimmers, and have a cruising speed of 10 miles (15 km) per hour, which they can keep up indefinitely. Thus a 15-year-old tunny might travel $1\frac{1}{2}$ million km in his lifetime. This roving life style is due to the fact that tunnies feed solely upon sardines, whose shoals they tirelessly pursue across every ocean.

Much sought after for the taste of their flesh, these fishes are the object of intensive fishing, which endangers their chances of breeding.

Detail of the prominent lips of the green wrasse (*Labrus turdus*).

A broken sea urchin offers a scanty meal to some rainbow wrasses.

A scorpion-fish swoops down on them; takes one of them in its mouth; turns around and swallows it greedily.

Tropical seas

In tropical seas during the day, several species of fishes can be seen gathering around coral massifs.

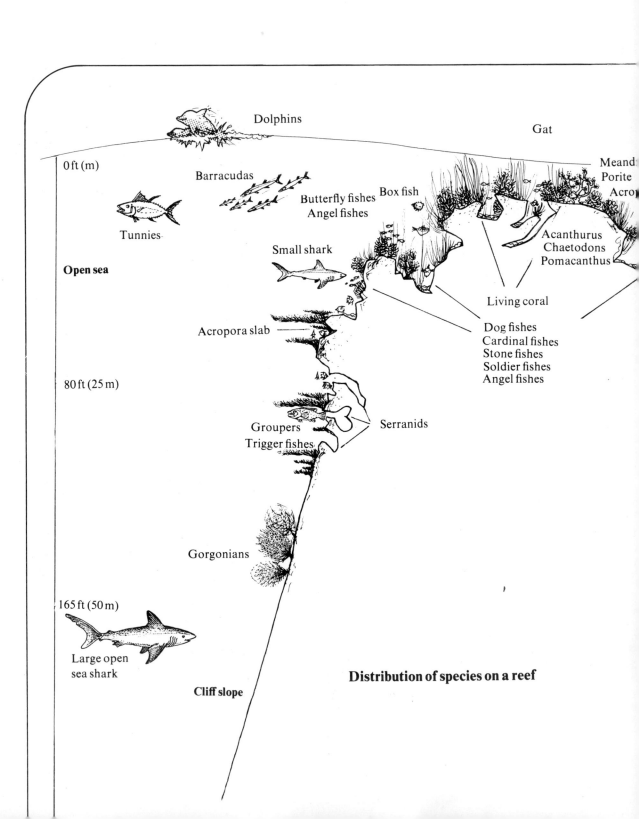

Dolphins

Gat

Mean[d]

Porite

Acro[?]

0 ft (m)

Barracudas

Butterfly fishes
Angel fishes

Box fish

Tunnies

Small shark

Acanthurus
Chaetodons
Pomacanthus

Open sea

Living coral

Acropora slab

Dog fishes
Cardinal fishes
Stone fishes
Soldier fishes
Angel fishes

80 ft (25 m)

Groupers
Trigger fishes

Serranids

Gorgonians

165 ft (50 m)

Large open
sea shark

Distribution of species on a reef

Cliff slope

I. The environment

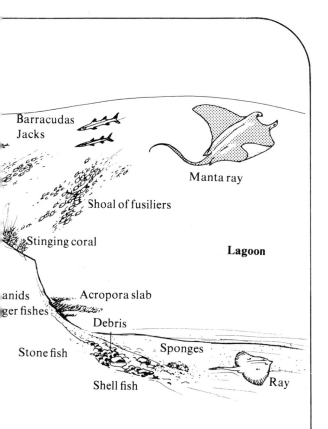

Barracudas
Jacks

Manta ray

Shoal of fusiliers

Stinging coral

Lagoon

anids
ger fishes

Acropora slab

Debris

Stone fish

Sponges

Shell fish

Ray

Tropical seas are in no way comparable to temperate seas. Among the factors governing them, their meteorological and biological cycles are basically conditioned by winds, currents and the clarity of the water. It is these factors that, on one hand, determine the atmospheric conditions, which are very uniform in character, and on the other the growth of animal life.

Winds

Along the equator, the temperature is elevated. Consequently, the air there is hot and rises, creating a depression in which the air masses originating from the tropics are engulfed. Hence the trade winds, steady winds always blowing in the direction of the equatorial tropics, are found both in the Northern and Southern hemispheres. Being a permanent feature, these winds establish a thermal regulator, bringing a refreshing coolness to coastal regions. As they are continuous in speed and direction, trade winds are a valuable ally to the navigator.

Monsoons are recurrent winds whose direction changes, and even reverses, depending on the season. In winter, the land, a high pressure zone, casts its icy air masses over the oceans; in summer, the overheated mainland generates low pressure and draws in the sea air. This account must of necessity remain very sketchy, as the mechanism is still not entirely clear. Monsoons are steady winds and are especially apparent in the Indian

Coral structures :
   Top : a coral barrier in Polynesia.
   Middle : an ireland surrounded by
fringing reefs.
   Bottom : a lagoon seen from the insid

Coral fishers in the Maldives irela
The natives use the coral as a sour
   lime. They are often to blame fo
      destruction of the

Ocean. From October to April, they blow from the Indian mainland towards the ocean right down to the Seychelles; from May to September, they settle in the opposite direction. However, this general pattern is subject to change here and there, depending on the geographical relief and cyclonic disturbances.

Finally, in tropical seas, huge whirling and ascending air movements, such as cyclones or hurricanes, can fiercely occur. These whirlwinds can both gyrate and move forward at great speeds. In the southern hemisphere, they gyrate in a clockwise direction; and in the northern hemisphere, in an anti-clockwise direction. Originating near the equator, they gradually move away along an east-west trajectory, sometimes attaining speeds of more than 120 miles (200 km) per hour. On nearing the tropics, they abate and then die out. Cyclones, although extremely dangerous, only wreak havoc in certain regions:—namely, Madagascar, the Comoro Islands, the seas around China and the Philippines, the West Indies, and the southern coast of America—and in well-defined meteorological conditions. They are accompanied by torrential rainfall and a residual swell which agitates the water for some days, and jeopardises diving.

Currents

Great ocean currents convey masses of warm or cold water, in such gigantic quantities that they can exert a strong influence not only on the climate of a country, but on its animal and plant life. In the northern Atlantic, the most famous of them is the Gulf Stream, which reheats every European coast as far as Norway. It originates from the north equatorial current, which flows in a westerly direction under the influence of the trade winds and drives enormous masses of warm water towards the Caribbean sea. In doing so, it promotes the growth of tropical species, coral reefs in

Ceylon.

particular. On the other hand, there is only a single temperate or sub-tropical zone in the southern Atlantic, which is swept by a cold current emanating from South Africa, moving up along Angola and flowing down again, having been only slightly reheated along the Brazilian and Argentinian Coasts.

In the Indian Ocean, a cold current runs along the western coasts of Australia. But it warms up again as it moves up towards the north and then bends down to the west. It then rejoins the south equatorial current along the East African coast, between the continent and North Madagascar, where it forms the warm Agullas current, one of the most powerful in the world. It descends along Mozambique and South Africa down to the tip of the Cape, where, despite the latitude, tropical species are still to be found. North of the equator, the Indian ocean has a much more confused system of currents, like that of an inland sea. Nevertheless, the temperature of the water is favourable for the growth of tropical fauna.

In the Pacific Ocean, the coastlines of California, Chile and Peru are washed by cold waters; in fact, the trade winds sweep the warmer surface waters towards the equator; coral reefs are only visible in the open seas of Central America. On the other hand, the west Pacific is swept by warm north and south equatorial currents: one flows down to Australia, the other rises towards Japan. In this way, tropical species can be spread over a zone of some 80° of latitude, even though they develop in a relatively restricted stretch along the American coast (about 30° of latitude).

Clarity of the water

The distribution of species in tropical water also depends on its luminosity, since the Zooxanthellae, those little algae which are symbiotically associated with corals, convert gaseous and mineral elements into organic products by photosynthesis. The maximum depth at which corals can live thus depends on the clarity of the water.

Thousands of silvery fish adorn the delicate coral branches on a sea bed which is bathed in light.

Gaterin orientalis along the Madagascan shores.

Stinging coral (*Millepora*). Its lovely branches are
deceptive: the least contact with them can result in
extremely unpleasant stings.

2. Invertebrates

Corals

Corals, madrepores, reefs or atolls are for many only a confused whole, bringing to mind warm climes, holidays, beaches and fabulous underwater fishes. Yet, more than all this, they form one of the finest and most considerable branches of the animal world.

Certain species actually constitute the prime element in the biological cycle and are felt to be an important factor in the geographical coastal configuration as well as in the creation of atolls and lagoons.

Corals can be either solitary or colonial organisms, essentially made up of polyps (the soft and living part). Although very primitive, these polyps are astonishingly well equipped both to defend and feed themselves by means of special cells like little egg-shaped capsules containing delicate spines attached to a filament which is rolled up and compressed like a spring. The whole thing lies floating freely in a stinging fluid. The least contact from outside involuntarily releases these minute harpoons, which can either paralyse and kill the prey, or sting and repel the predator.

In some instances, the polyps secrete a calcareous or chitinous skeleton (for example, gorgonians). To be more precise, there are specialised cells secreting a vital organic substance which serves as a matrix for aragonite mineral fibres (calcium carbonate). The limey secretion is not the same at night as it is in the day. This seemingly unimportant point has

actually proved of cardinal importance to the scientific world. In fact, this varying growth rate is shown by minute excrescences which are witnesses to the diurnal cycle of the earth. A simple comparison between coral fossils of 400 million years ago with present-day corals has shown that in that era, a year consisted of 400 days, and that consequently days are getting shorter by 1 second every 132 years.

The structure of these animals, which is extremely complex and architecturally flawless, gives remarkable solidity to each construction unit: calcareous strips, a columella, and horizontal laminations form a particularly rigid cohesion.

Being mostly hermaphroditic in nature, corals reproduce by means of specialised cells. Rudimentary testes and ovaries can be found either on the same polyp or on different polyps in the same colony. Spermatozoa and ovules produce an egg, which grows on the polyp itself, and gives birth a short time afterwards to a little ciliated larva. This leads a free planktonic life for several days, before attaching itself to a suitable substratum. The larva is then transformed into a polyp, in the colonial species, which buds and creates new polyps, which will bud in their turn, so enabling the colony to build those magnificent edifices.

There are other reproductive systems such as active buds, which break loose in order to lead an autonomous life; and the apportioning of the skeleton itself into component parts, which are capable of reproducing themselves.

Polyps withdraw into the skeleton during the day, and bloom out in the

Soft coral (Alcyonarian spongode without spicules) in the Red Sea: it has an infinite variety of colours.

Soft coral (*Sarcophyton trocheliophorum*), with its polyps retracted

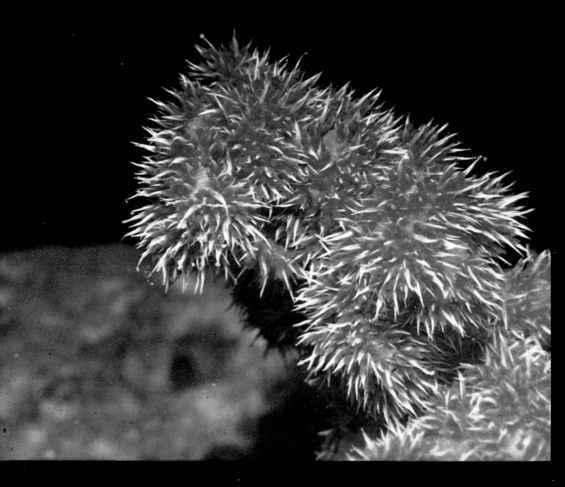

Spiculiferous alcyonarian
found on the giant reefs of
the Indian Ocean. Although
its body is soft, its tips bristle
with prickles.

Pacific alcyonarian of the
Morchellana genus. It has
spicules like a sponge.

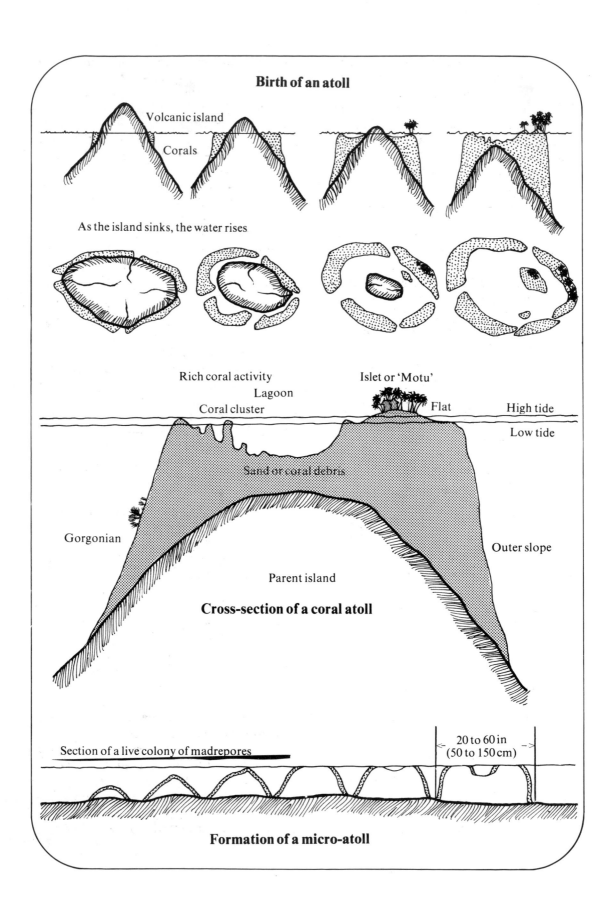

Birth of an atoll

Volcanic island

Corals

As the island sinks, the water rises

Rich coral activity

Lagoon

Coral cluster

Islet or 'Motu'

Flat

High tide

Low tide

Sand or coral debris

Gorgonian

Outer slope

Parent island

Cross-section of a coral atoll

Section of a live colony of madrepores

20 to 60 in
(50 to 150 cm)

Formation of a micro-atoll

evening in such a way as to stir up a large volume of water and capture prey, which will then be swallowed and digested.

The growth of corals was for a long time the subject of controversy. It was noticed that microscopic algae (dinoflagellates) were found in their tissues. These 'algae' contain chlorophyll, and are thus midway between the plant and animal kingdoms. It is apparent that they live in symbiosis with the polyps. They take advantage of their calcite shelter, but in exchange for certain chemical components that corals provide them with, they make organic substances which enable corals to fix their calcium.

Their rate of growth varies from a fraction of an inch to several inches (about 40) a year. It depends on the age of the colony, which, contrary to popular belief, is not everlasting; young polyps do not use their parents as a calcareous support indefinitely. In reality, their life expectancy rarely exceeds a hundred years.

Many tropical islands are surrounded by circular reefs from which marvellous lagoons, sheltered from the ocean swell, are formed. The formation of an atoll is often due to a volcanic isle progressively subsiding or, sometimes even, to a rise in the sea level. It can develop in different ways:

1. The volcanic island is young and completely emerged from the sea. A coral fringe slowly grows in those places most suited to coral species. The river mouths alone remain unencumbered because corals do not tolerate fresh water very well (e.g. the Grand Comoro).

2. As the island sinks, madrepores expand upwards in order to maintain their usual conditions of life: depth, luminosity, etc. Some species which have adapted to living water find themselves in stagnant water and die; little lagoons as well as channels start to form at the mouths of rivers. As can be seen on Reunion, the channels allow the lagoons to empty or fill following the tides.

3. The island is almost submerged. The lagoon has been formed and some of the channels are blocked up because the rivers have disappeared. There only remain one or two channels which allow the water levels to change with the varying tides. Coral reefs ('ves motus' in Polynesian) emerge from the sea forming little islands, like Bora Bora.

4. The island has completely vanished, an atoll has been born. Sometimes it is enormous, like Rangiroa, which is 112 miles (180 km) long. Usually a sizeable channel continues to exist, although there are totally enclosed atolls which continue to maintain an ever-changing fauna.

The majority of corals live all their lives attached to a substratum, except those bound by a small peduncle which have freed themselves during their growth in order to live a free and mobile life. The mushroom corals (*Fungia*), for instance, can, by contracting their polyp, go backwards and forwards to avoid being sucked down into the crumbly sand.

Other free madrepores use very artful systems: for example, the *Heteropsammia* contains a tiny parasitic worm whose movements reverberate throughout the whole colony. Another species has kinds of flexible ballasts which, when alternately filled with water, first on one side then on the other, enable the colony to advance or retreat.

Stinging corals

These hydrozoans are so-called because they possess a gastroenteric sac armed with formidable stinging tentacles. They can either be devoid of a skeleton, like the water hydras, or be provided with one, like the chitinous *Solanderia* and the calcareous *Millepora*. Unlike the *Solanderia*, the *Millepora* can live perfectly well near the surface and resist the sudden overflow of spring tides. The bad reputation of the stinging corals is not without cause. The least contact can result in extremely unpleasant, although not serious, burns or skin complaints.

Soft corals

On the other hand, there is nothing to fear from soft corals (Alcyonacea, Anthozoa). They have no skeleton, possess eight tentacles—or a multiple of eight —and two polyps, a digestive one and a predatory one. Among these various corals, spongodes live on the outer slope of reefs at a depth of at least 30 ft (10 m) and never surface. Sarcophytons themselves prefer to frequent those parts buffeted by the waves.

Gorgonians

Contrary to appearances, a gorgonian (*Gorgonacea*) does possess a skeleton. But this often so resembles, and is as supple as, a plant spray that it is easy to make this mistake. Gorgonians reach considerable sizes: recently a gigantic specimen was caught off Reunion, measuring more than 10 ft (3 m) high and 8 ft (2·50 m) wide. They always settle at an angle perpendicular to the current, with a view to culling the greatest amount of food. Like the alcyonaceans, they are able to live at various depths, extending from the submerged parts of the reefs down to several hundreds of feet. From these animals is extracted an organic product that is both strong and pliable: this is gorgonite or 'black coral'.

The purpose of the gorgonians and alcyonaceans in the biological cycle has not yet been elucidated, owing to the lack of information about their physiology and metabolism. All the same, it is known that they live on reefs, contain algae (*Zooxanthellae*) and that both fishes and shellfishes browse off some species. Moreover, it appears that a kind of cement which slowly fills the cavities of the reef originates from the dead bodies of Alcyonaceans.

Common corals

Taking up nearly 80 per cent of the surfaces of reefs, these animals are characterised by six (or a multiple of six) tentacles. Most numerous of the common corals are the *Acropora*, which are

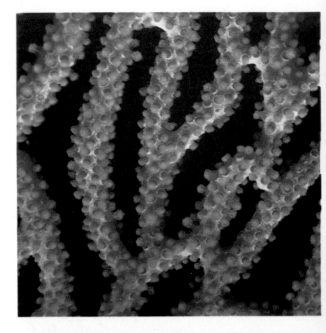

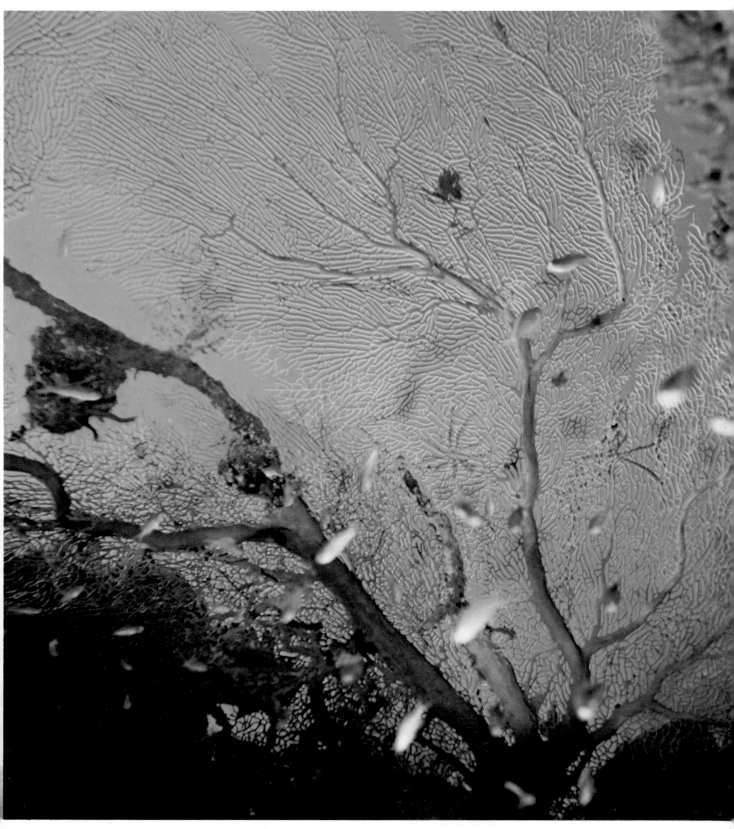

Polyps of a gorgonian.

Fan-shaped tropical gorgonian. Some are gigantic
and can measure up to 10 ft. (3 metres) in length and
7 ft. [2 m] in breadth.

Slab of stagshorn coral (*Acropora*). These slabs are so fragile that they are only found in well-sheltered areas.

Above: Usual appearance of a mushroom coral
(*Fungia*).

Below: The polyps of these corals (*Goniopora*) only
open out during the day.

Above: Detail of the calcareous structure
of a mushroom coral.

Below: Usual appearance of a *Goniopora*
coral.

85

themselves divided into a great variety of species. Their proliferation is explained by highly developed faculties of adaptation which enable them to live on almost every substance, at all sorts of depths, needing neither too much light nor too well sheltered surroundings.

Furthermore, as the *acropora* grow at a very rapid rate, it is not surprising that their structure is often extremely large, and that their development takes various forms: encrusted, small and hidden when they fix on to a substratum; branched, bushy and widely expanded in calm zones; or digitated and stocky in the troubled zones of the reef; or even in impressive slab-like shapes, which in some specimens can reach 16 ft (5 m) in length.

Mushroom corals (*Fungia*), are fairly common in tropical seas, where they form free-living colonies. They have the ability to move about, for when they become adults, the peduncule which clings to the substratum can no longer support their weight, and breaks. The polyp is particularly active; it contains vibrating cilia which enable it to travel over 8 in (20 cm) in a night, and so prevent it from sinking into the sand. They can also be identified by their single aperture.

The *Goniopora* are distinguished by their widely expanded polyps, which they leave during the day—an uncommon occurrence in the coral world. Furthermore, they sometimes break loose from their support and fall onto the sea bed, where the colony then forms itself into a very odd-looking ball.

Top left: Detail of the tip of a gorgonian branch, showing the position of the polyps on their horny skeleton.

A forest of gorgonian polyps in the Indian Ocean.

Opposite: Detail of the calcareous structure of the casing of polyps of *Acropora* corals.

Composition of a slab of
Acropora corals in the
juvenile stage. The branches
multiply from a stalk which
is firmly fused to the sea-
bed.

Detail of the calcareous
structure of the casing of
polyps of the *Seriatopora*
genus, shaped like antlers.

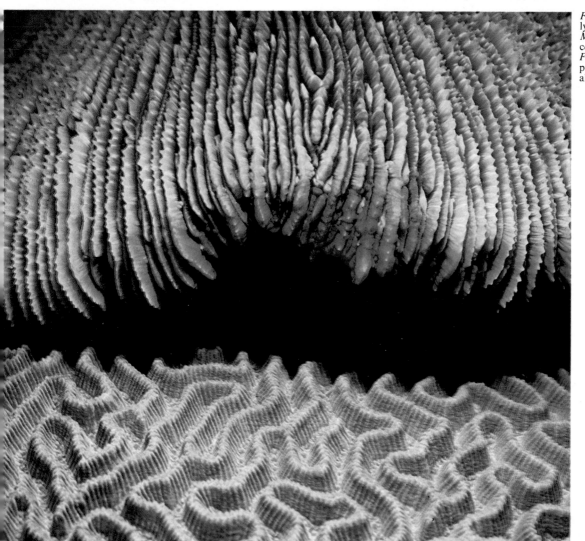

Fungia: Mushroom coral
lying over a coral of the
Meandrina genus. The violet
colour of the base of the
Fungia is caused by the
presence of microscopic
algae.

Molluscs

Molluscs play a primordial role in the animal world. They are scattered throughout every corner of the world and have been able to adapt themselves to most conditions of life on earth: high altitudes 20 000 ft (6000 m), unfathomable depths, fresh water, cold water, hot water, etc. They are characterised by certain typical features, such as the muscular 'foot' which is often very powerful and enables the creature either to crawl or to cling to a rock wall; the 'mantle', a kind of membrane, often richly decorated, which secretes a calcareous skeleton; finally, the 'radula', a basic element in the study and identification of the species, which is a sort of raspy tongue used either to crumble or grind ingested food into fine particles.

Monoplacophores

At the present time, six species of monoplacophores have been indexed. They comprise a group recently discovered (1952) by the Danish expedition in the vessel *Galathea*. Some examples of a small shellfish looking like a limpet were brought up from a depth of about 12 000 ft (3500 m). They were named *Neopilina galathea*, and were shown to be descendants of a branch that were thought to be extinct for some 350 million years, and which had the characteristic, unique in molluscs, of possessing double organs.

Chitons

Generally known as chitons, the Amphineura class is easily distinguished from other classes of molluscs by the shell consisting of 8 plates, each of which articulates with the next, enabling the animal to roll itself into a ball, like a little woodlouse. They are very common in all oceans and are often on friendly terms with limpets, preferring the shallow coastal areas. The most widely spread species is the *Cryptochiton stelleri*, which can grow to 12 in (30 cm) on the western shores of the United States and Canada. Its plates are not usually visible, as they are covered by a hard skin.

Cowries

These molluscs are part of the Gastropoda class, like turbans, ceriths, ovules, naticas, conches, tritons, murexes, olives, cones and sea slugs. This class is characterised by a helical shell. In some cases, this feature diminishes during growth and becomes almost invisible, notably in the limpets and some cowries. In general, gasteropods move forward by creeping along on their foot, which contracts and then expands after secreting a mucus. This temporarily cements together little particles and grains of sand to form a veritable carpet, over which the animal slides. During the embryonic period, the body of the mollusc undergoes a remarkable torsion of 180°. This phenomenon has the effect of displacing the respiratory organs which, in most cases, stay in this position. This is typical of the sub-class Prosobranchia, which includes most known shellfishes, with the exception of sea slugs.

The genus *Cypraea* comprises about 190 species, grouped under the name 'cowries'. For a long time, these shellfishes have fascinated man, either as collectors' items, decorative objects, or subjects of scientific interest.

Their structure and the final aspect of their shell make them an exception among shellfishes. In fact, in 99 per cent of cases, with gasteropods at least, shellfishes are coil shaped with an aperture at one end, which is set off at an angle from the axis of the spiral. With cowries, on the other hand, the last turn of the spiral is distorted by callouses, which make it look like a tortoise-shell, and the aperture opens out in the axis.

Growth is rapid in the first stage of development, and the animal reaches its maximum size in two or three months. Being still very smooth and delicate, the shell then thickens. The apex disappears, teeth—labial, and then columellar —develop, and two labial callouses appear. Finally, the last layer of shell, incorporating the design is formed. The animal is then able to dissolve part of its inner strata in order to increase the size of its living quarters.

Diagram of a gasteropod

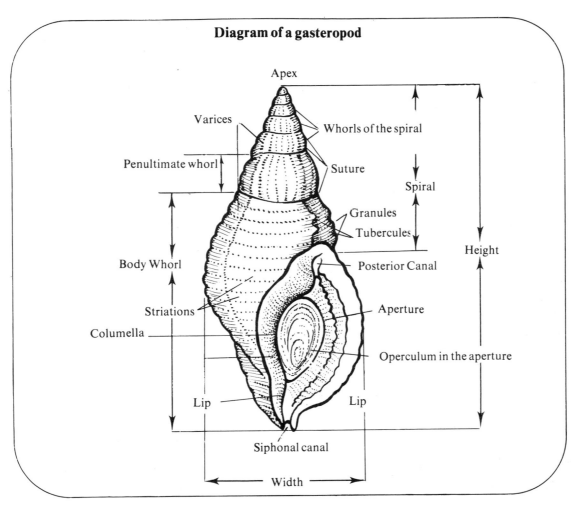

Apex

Varices

Penultimate whorl

Whorls of the spiral

Suture

Spiral

Granules

Tubercules

Height

Body Whorl

Posterior Canal

Striations

Aperture

Columella

Operculum in the aperture

Lip

Lip

Siphonal canal

Width

Close up of a cowrie turning over.

A – It is laid on its side and the mantle, having emerged from the shell, now covers it completely.

B – The cowrie is upright again. Its eyes can be seen on top of the antennae.

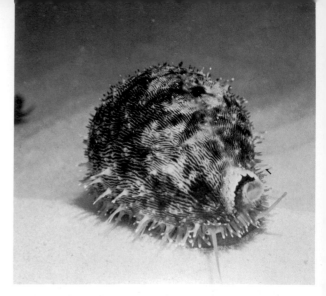

Protection is afforded by the mantle, which ensures the superb sheen of the shell. It covers the shell totally and permanently, preventing any animal or plant parasites from attaching themselves to it and spoiling it. The mantle is made up of innumerable papillae and probably also acts as a supplementary respiratory organ. In some cases, its bright and varied colours can create effective camouflage.

Cowries are, without exception, herbivorous. They feed on minute algae which they pound on the substratum with their radula. Many species also graze on sponges, some being carniverous as well.

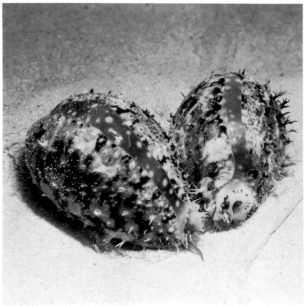

Cowries are found throughout the world. Although the majority of them live within tropical bounds, a certain number of species, however, have become adapted to temperate waters. They are found as far afield as South Africa, around latitude 35° south, as well as in the Mediterranean (*Cypraea pyrum* and *C. lurida*), at a latitude of 45° north.

The tiger cowrie (*Cypraea tigris*) is certainly the most common of all, and is often the first item in a collection of shells. They are polymorphous, with a pigmentation ranging from milky white to dark blue. They measure 1 to 6 in (3 to 15 cm) (only in outstanding specimens belonging to the Hawaiian *Schilderiana* variety) when they are fully grown.

The deer cowrie (*Cypraea vitellus*) varies in size from 1 to 3 in (26 to 80 mm) and is often confused with the camel cowrie (*C. camelopardalis*). The latter can be differentiated by visible gaps in its black teeth. Moreover, they do not live in the same regions: the 'deers' are found in the Indo-Pacific zone, the 'camels' in the Red Sea. The *Cypraea nivosa* resembles them both, but is much rarer, having a distinctive greenish colour. The carnelian cowrie (*Cypraea carneola*) also figures among the species common in the Indo-Pacific region. Both its shell and its mouth are violet, so enabling it to be distinguished from the *Schilderorum* and the *Sulcidentata*. However, seen from the underside, they are very similar. The car-

Top: Tiger cowrie (*Cypraea tigris*): one of the most widely distributed species in almost all tropical seas.

Middle: The deer cowrie (*Cypraea vitellus*), so called because of the little white dots scattered over its brown shell.

Below: Carneola cowrie: it is brightly coloured when alive, but its colours grow blurred once it is taken out of the water.

90

'The black pearl of the Red Sea' (*Cypraea exusta*) is found in the southern part of the Red Sea.

Cowrie (*Cypraea nebrites* in the adult stage). It can be distinguished from the *Cypraea erosa* by large black patches on its lateral calluses.

Nucleus cowrie (*Cypraea nucleus*). The mantle covers its shell and serves as a camoflage when it changes position.

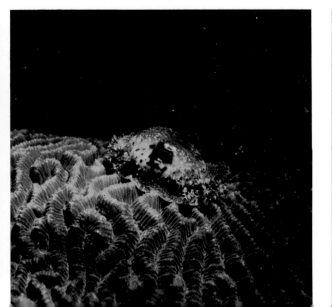

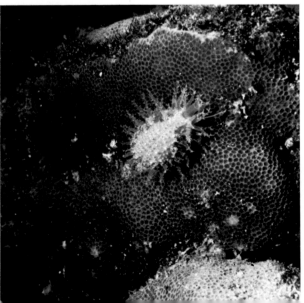

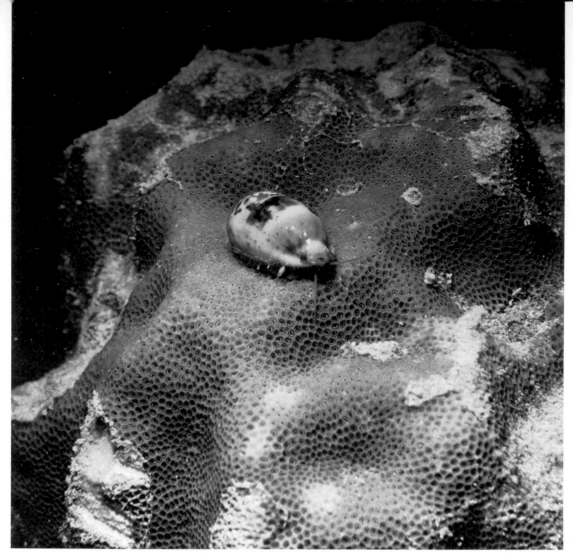

Red-spotted cowrie: a rare cowrie which was for some
while found along the coasts of Ethiopia.

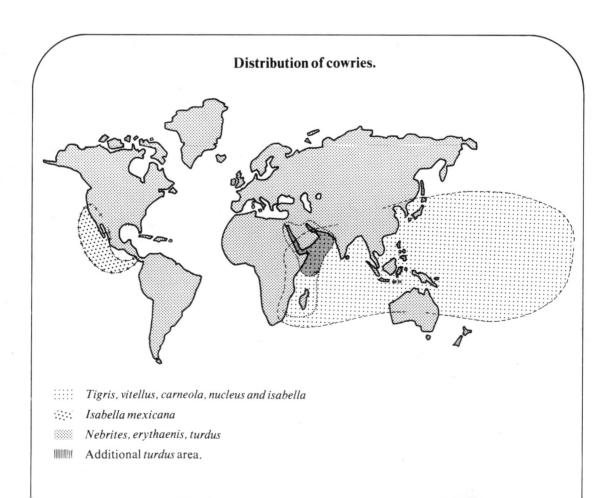

Distribution of cowries.

::::: *Tigris, vitellus, carneola, nucleus and isabella*

:·:·: *Isabella mexicana*

::::: *Nebrites, erythaenis, turdus*

‖‖‖‖‖ Additional *turdus* area.

nelian cowrie, like the tiger, comes out to feed during the day.

The *Cypraea nebrites*, which lives in the Red Sea and its surroundings, is characterised by its very open labial callouses. It scarcely exceeds 1–1½ in (25–35 mm) in size. With its two black labial markings, it can be mistaken for the *C. erosa*. It can only be identified by examining the base; when seen in this position, the markings of the *nebrites* are hidden, whereas part of those of the *erosa* are visible.

The nucleus cowrie *(C. nucleus)* is the main representative of the rough-surfaced species. It differs from the coarse-grained species in its small size—½–1 in (15–28 mm). The papillae of its mantle are exceptionally long.

The Isabella cowrie (*C. isabella*) is one of the most beautiful sea-shells. Decked out in a jet black mantle which highlights to perfection the brilliance of its orangey-red extremities and its white undersurface, it is without equal in the underwater world, save for its close cousin, the *Isabella mexicana* which is found on the western coast of Central America. Their

one difference is visible on the under-surface, which in the latter is a creamy colour. Their dimensions, however, are similar: ½ to 2 in (11 to 54 mm).

The thrush cowrie (*C. turdus*) is very common in the Red Sea, but rare specimens can also be found on the Kenyan and Madagascan coasts. It measures between ¾ and 2 in (20 and 48 mm) with wide labial callouses (notably those found in the Gulf of Tadjura), and appears to have remarkable immunity to fishes when it wanders, preferably during the day, over the living reefs.

The red-spotted cowrie (*C. eryth-raensis*) is small in size ⅔–1 in (16–27 mm), and has for a long while been considered as something of a treasure by collectors, in view of its rarity. But the recent discovery of several colonies along the shores of Arabia and the areas around Djibuti has brought down its price. It was originally placed in the sub-order Stoli-dobranchiata, from which it has now been excluded because of numerous morphological differences. It is appreciably smaller and more bluish in colour.

Thrush cowrie (*Cypraea turdus*), very common along the coasts of the Red Sea. It also sporadically appears along the shores of Kenya.

Isabella cowrie *(Cypraea isabella),* characterised by a black mantle and two little red patches at the tips of the apertures of its shell.

Warty ovules

These are represented in this book by the *Calpurnus verrucosus*, a delightful little shell which does not merit its name. Apart from its two pinkish tips, its shell is completely ivory coloured. It is distinguished from true ovules by its bulkier shell, which is both humped and squat, and by two little warts placed around the tips in two small subterminal depressions. It is the only ovule to have teeth on the labial lip. The *Calpurnus verrucosus* lives on certain of soft corals, on which it feeds.

Turbans

These make up the order Archaeogastropoda consisting of about 500 species, the most spectacular of which live on coral reefs. They differ from the *Trochidae* —which are remarkably similar to the *Astraca* genus—in having a very solid calcareous operculum. Many species of turbans live on barrier reefs or in zones with adjoining promontaries, and as a general rule wherever considerable hydrodynamic forces are at work. They feed by grazing on algae, and thus prevent them from invading the stagnant shores of coral colonies. The shells of some turbans are used to make buttons.

Ceriths

There are more than 250 species of ceriths which have overrun areas as varied as shallow sandy borders, lagoon beds, reefs of coral islands, or even mangrove swamps. They are herbivorous or detritivorous, and belong to either the epibiotic or endobiotic groups of fauna. The order is characterised by a great variety of sizes, colours, and even designs.

Naticas

These are carnivorous necklace-shells which live in the sedimentary beds of nearly all seas in the world. Their mantles have wide lobes covering most of their shell, like those of the cowrie or the ovule. They feed on other molluscs, whose shell they pierce, employing both their radula and acidic secretions. The ovigenous capsules of the naticas are arranged in long spiral ribbons, making their spawn oddly collar-shaped.

Conches

This family brings together over 100 tropical species with shells that are thick, ellipsoidal, tightly folded, and pitted with more or less regularly spaced excrescences, which indicate the successive stages of growth.

Unfortunately, owing to over-intensive fishing, conches are fast disappearing. It is this above all that is responsible for the disastrous proliferation of those starfishes known as 'crown of thorns', in the Indo-Pacific region, which feed on corals, and do increasing damage, particularly on the Great Barrier Reef and on Guam. Conches are the only mollusc predators of these starfishes.

Murexes

These are also carnivorous molluscs. Their shells are often provided with quite a long pitted tail and spiny varices, which can be either foliated or tightly compressed. Drupes, which belong to this family, are either herbivorous or carnivorous; they abound on the ridges covered with red limey algae of the coral reefs of the intertropical zone.

Murexes, which number 800 species in their group, are spread throughout every ocean in the world. They live, like the *Murex anguliferus Lamarck* of the Atlantic, on rather muddy sedimentary beds strewn with coral debris.

The female of the species lays a large number of ovigenous capsules clustered together in an odd spawn, which is shaped like the animal itself.

Most of the murexes feed on their own species or on other molluscs, which they attack from the rear perforating their shells with their powerful radulas.

Olives

Although the Olividae family only comprise about 60 species, nearly 200 varieties of shells with different polychromatic patterns can be identified.

The glossy, almost cylindrical olive has a short spire whose whorls are separated by a little groove in which a fibrous appendage of the covering mantle is embedded.

Olives are carnivores, living in the coralline or volcanic sands of tropical shores. They can be captured with bait.

Cones

These are an extremely numerous (about 500 species) and varied family. They are dangerous predators, armed with a small harpoon which they skilfully use to attack little fishes and defend themselves against man. It is advisable to beware of them! Some cones, such as the *Conus geographus, C. textile, C. tulipa, C. amaria, C. bullatus*, secrete a toxin which can prove fatal. It is stored in a sac, the leiblin gland, and emitted through the pharynx where it permeates types of darts which are in fact oblong-shaped detachable radulas with barbed and pointed tips. They are assimilated in the radular sac which contains twenty of them. One by one they are drawn out and brought up through the pharynx, from where they are taken to the proboscis and propelled into the bodies of their victims. Their trajectory can reach an equal length to that of their shell. In the dart, a longitudinal groove enables the toxin to be simultaneously injected.

Although not all cones are so formidable, many can nevertheless inflict pain, not to mention localised and temporary paralysis. It is therefore imperative to be wary when collecting them: to pick them up by the undersurface and never to hold them in the palm of the hand, but to immediately place them in a bag.

Because of their pronounced polymorphism, classification of cones is a problem that arouses much controversy. Some people think that they constitute a single genus. The Natural History Museum in Paris distinguishes four genera, each divided into 25 sub-groups.

The *Conus geographus* has a relatively

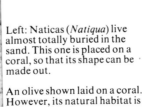

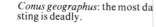

Conus geographus: the most da[...] sting is deadly.

Conus sumatrensis: often found [...] of the Red Sea.

Left: Warty ovule (*Calpurnus verrucosus*), often found during the day clinging beneath corals of the *Sarcophyton* genus.

Naticas lay their eggs in a shape like that of a suspension spring in a mixture of sand and mucus which they produce. ▶ ▶

Left: Turbans have a brightly coloured calcareous operculum, which is highly prized as jewellery.

Spiny murex of the Red Sea laying its eggs. ▶ ▶

Left: Ceriths (*Cerithium fasciatium: Terebra*) always live embedded in the sand – they can only be located by following their trail.

Murexes are cannibals! A murex is here seen 'drilling' through another murex with its radula in order to pierce its shell and devour it. ▶ ▶

Left: Naticas (*Natiqua*) live almost totally buried in the sand. This one is placed on a coral, so that its shape can be made out.

An olive shown laid on a coral. However, its natural habitat is in the sand. ▶ ▶

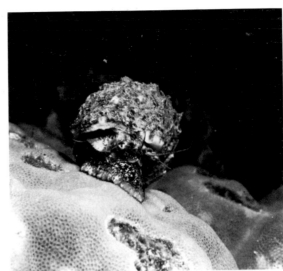

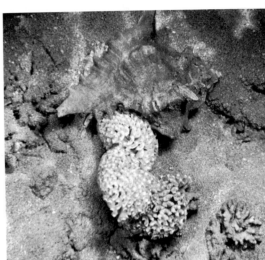

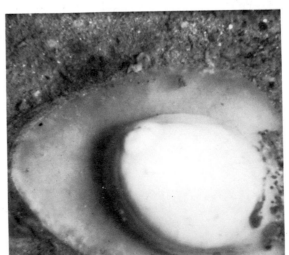

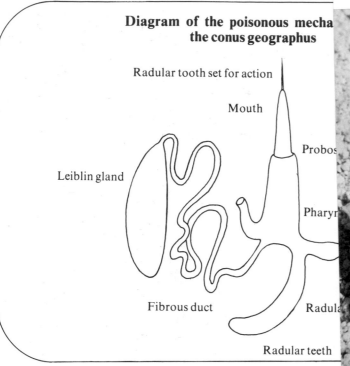

Diagram of the poisonous mecha...
the conus geographus

Radular tooth set for action

Mouth

Probos...

Pharyn...

Leiblin gland

Fibrous duct

Radula...

Radular teeth

The *Conus textile* belongs to an extremely dangerous
group of death-dealing cones. It uses its dart to
harpoon little fishes on which it feeds.

The *Conus cuvieri* is a little coral cone living along the
reefs.

The Molluscs

PHYLUM	CLASS	SUB-CLASS	ORD...
Mollusca	Gastropoda	Prosobranchia	Archa... Mesog... Stenog...
		Opisthobranchia	Pleuro... Pterop... Sacog... Acoela...
		Pulmonata	Basom... Stylom...
	Amphineura		Lepido... Chiton... Neome... Chaeto...
	Monoplacophora		Tryblic...
	Scaphopoda		
	Bivalvia		Protob... Filibra... Eulame... Septibr...
	Cephalopoda		Tetrab... Dibran...

The *Conus terebra* is commonly known
as the Auger shell because of its
elongated appearance.

A sand cone (*Conus arenatus*) usually completely
buried some (centimetres) under the sand. It can be
traced by the tracks it leaves on the sea bed.

Conus textile laying its eggs.

Striate cone (*Conus striatus*).
Particularly prevalent in the coral
massifs of the Indo-Pacific region.
It lays its eggs in the shape of a small
pouch under stones or dead corals.

light bulb-shaped shell. It can reach a size of 6 in (15 cm) and its aperture is very wide. It is found in the Indian Ocean and in the Pacific, but it is especially common in the north-west Pacific (the Philippines). Pre-eminently a predator, it only eats fishes, and lives in hiding in sandy zones, under clusters of corals.

The *Conus textile* is a pretty, almost concave, polymorphous cone, with a moderately raised whorl. The dozens of names attributed to it seem only to be synonyms or ecological varieties. It is distinguished from most other species by black zig-zag lines running through large maroon patches. It is also a dangerous predator capable of causing death. In exceptional cases, it can reach a size of 5 in (130 mm) and lives throughout the Indo-Pacific zone in sandy beds, beneath corals and rocks.

The *Conus arenatus* is a common diurnal species, characterised by the thousands of dots which adorn its shell. It is somewhat concave in shape, with a coronated spire and its size does not exceed 3 in (75 mm). It is found in every Indo-Pacific area, living in sandy beds, where it leaves a typical trail.

The shell of the *Conus terebrus* is large, elongated, heavy and white, sometimes intersected by pink, yellow or violet transverse bands. Its surface is generally wrinkled by little cross grooves, except in some specimens when it is quite smooth. The *Conus terebrus* is encased in a thick layer of chitinous matter (the periostracum) and does not appear to be alive. It reaches a size of 4 in (10 cm) and lives in the Indo-Pacific zone.

The *Conus cuvieri*, although little known, is frequently encountered in the southern zone of the Red Sea. Its dimensions can vary from $\frac{3}{4}$ to $1\frac{1}{2}$ in (20 to 40 mm). Its light, concave shell has a very wide aperture and ranges in colour from greenish-grey to brownish-black. It is fond of madreporian sanctuaries.

The striate cone (*Conus striatus*) has a beautiful, heavy, almost cylindrical shell with a low spire. As its name implies, it is evenly striped over the total surface area of its last whorl. Its ornamentation consists of quite significant large grey patches on a white or pink background. Specimens collected from the Red Sea often have simple semi-triangular shaped designs. They reach a length of 4 in (10 cm) and are encountered in the Indo-Pacific area. They are avid predators, living on little fishes, and having the reputation of being dangerous to man.

Sea slugs

These aquatic snails differ from the Prosobranchia in that their respiratory organ is located behind the heart. This group of gasteropods is subdivided into two main branches: the Pleurocoela, and the Nudibranchia. In the former, the gills are protected by a much reduced shell. Consequently, in the sea-hare (*Aplysia*), it is scarcely perceptible through the siphon. Occasionally, when it is completely covered by the mantle, it is not even visible. In the latter, the naked gills are on the back—the shell having completely vanished. The respiratory apparatus can be located at the rear of the body, like a star-shaped rosette surrounding the anus, as in the Doris; or it can be made up of numerous cerata arranged on both sides of the body, as in the Aeolids. These respiratory cerata can be either elliptical or grouped in clusters.

Of the two branches, the most spectacular is the Nudibranchia, because it includes the most graceful and colourful members. These invertebrates crawl along on a muscular pad-like foot, and they can change their shape at will. The front part of the animal, representing the head, has two olfactory tentacles (rhinophores) attached to it, which are the same colour as the gills. The body is decorated with motifs in vivid colours.

These adornments can be stripes, spots, or patches. They can assume the most bizarre shapes and the most varied hues, especially in the tropical species. Unfortunately, even fixing agents cannot hold these colours. Only a sketch or a

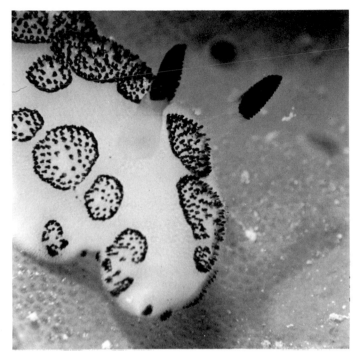

Spawn of an Opistobranch.   Head of a Nudibranch (*Glossodoris*).

Horny sea slug (*Phyllidia bourguini*) found all over the Indian Ocean.   Nudibranch (*Ardridoris tuberculata*).

A lavishly ornate Red Sea slug (*Phyllidia sp.*) lying on
an alcyonarian of the *Sarcophyton* genus.

A brightly coloured Red Sea slug (*Glossodoris*): its
gills are placed on its back like a little feather-duster.

colour photograph can give some idea of the brilliance of these animals.

The Doris usually feed on sponges, in which they dwell. When they spawn, attached to a support (rocks, sponges, algae), they lay thousands of eggs in long ribbons.

The Aeolids browse on colonies of hydroids, which are animals possessing stinging cells, like the sea-anemones. A curious phenomenon is that they store the stinging cells in the tips of their cerata in special little pockets (cnidosacs) and they subsequently use them to defend themselves. When the stockpile of little darts is exhausted, they renew it by devouring other hydroids, using their radulas, the rasping mechanisms contained in the buccal bulb. Their spawn is stringy and coiled, usually displaying the dominant colour of the body of the particular nudibranch.

These species are extremely numerous in the Mediterranean, whose luxuriant beds with their coral structures and grottos are the favourite places of these delicate creatures that can only be collected by diving for them. Even then, sponges, algae, and hydroids must all be meticulously observed, as the Nudibranchia are generally small in size—from a fraction of an inch to 4 inches (a few millimetres to 12cm) with the exception of some tropical species, and can easily be overlooked by the inadvertent diver.

Strombs

The Strombacea family comprises four genera: *Lambis, Terebellum, Tibia* and *Rimella*, which are all, except the last-named, well known to collectors.

Although descended from a considerable fossil group, today they are no longer very numerous. The bubonius stromb, common in its fossilised state, no longer very numerous. The *Bubonius* has been reduced to seven species in the Atlantic and to sixty in the Indo-Pacific oceans.

Strombs mostly live in the shallows. They are herbivorous and detritivorous.

They possess a very fragile foot, at the tip of which is a long pointed operculum which it uses both to move and defend itself.

Their eyes are always very highly coloured and are located at the ends of two long peduncles. This gives them greater mobility. Identification of the males, which are often smaller, is generally facilitated by their sexual dimorphism.

The Strombacea are easy to distinguish from other families on account of the slight depression (known as the stromboidal cavity) on the anterior part of the lip. One of its two eyes makes use of this depression, and the other uses the siphonal cavity, to keep watch over its surroundings. Nothing else of the shell shows when it is lying flat on the sand.

Strombs regularly lay eggs, joined together by a gelatinous substance, and sometimes rolled up in a spiral as do the *Tibia* species of Djibuti. About 80 to 100 hours later, little swimming larvae, which soon secrete the basic substances for the formation of their shells, are born.

Pelecypods

This class of molluscs groups together the bivalves in the two sub-classes: Lamellibranchia and Protobranchia. Mussels, oysters and scallops form part of the Pelecypoda class.

Although often ignored by collectors, they are most interesting shellfishes.

Some of them are historically very famous. Giant clams are the largest shellfishes in existence, the giant clam (*Tridacna gigas*) in particular: a specimen in the collection of the Museum of Natural History of New York weighs 579·5lb (about 260kg); another, caught off Sumatra, is 54in (137cm) long. France has a famous giant clam in the church of Saint-Sulpice at Paris, which weighs 595lb (270kg) and measures about 3ft (1m) across. François I received it from the Venetian Republic in the sixteenth century.

Giant clams have other surprising characteristics; to begin with, they cultivate small algae, which are then used as food, within their own shells. Their beautiful mantle comprises hundreds of little alveoles through which solar energy is infused, giving rise to photosynthesis and the subsequent growth of these algae. This cultivation is so important that, after some time, the giant clam feeds on nothing else.

Giant clams as they remain half open in order to expose their mantles to the sun, have earned a bad reputation of snapping shut on the feet of those swimming near them.

In fact, this is pure legend, as the aperture is relatively narrow and its closing powers quite slow, especially in the large specimens, which have to drain off dozens of litres of water beforehand.

Over the centuries, hundreds of men have dived and died fishing for oyster pearls. Pears are produced by chance: a little piece of debris or a grain of sand or nacre slips under the mantle of the creature, which, as a reaction, tries to smother what it regards as a parasite by enveloping it in nacre. Once in 10 000 times, a perfect pearl will emerge from it.

For some years, man has learnt to copy nature by slipping a little piece of nacre into cultured oysters, which do the same work, but with a surer yield. It is very dif-ficult to distinguish a cultured pearl from a real one, and sometimes only by observing the centre of the pearl through a hole can the piece of nacre specially implanted be differentiated from the accidental grain of sand.

Long ignored owing to the massive advent of cultured pearls, real pearls are today in greater demand than ever, and pearl fishing still continues in some parts of the world.

In Polynesia, oysters are fished for their fine black sheen, which is very highly prized in marquetry, in the 'curios' of Tahiti or New Caledonia.

Annelids

These segmented worms make up the phylum Annelida. One of their group, the polychaetes, are often mistaken for plants or underwater flowers on account of their helical umbrella-like appearance. This mistake can prove unpleasant for swimmers. Some of the annelids hide in their byssus fine calcareous spines capable of causing very painful stinging injuries together with numbness. One of them, the *Lumbriconereis heteropode*, has the further disadvantage of occasionally inducing headaches and vomiting.

Porites.
A colourful group of the *Sabellidae* family seen lying on a coral.

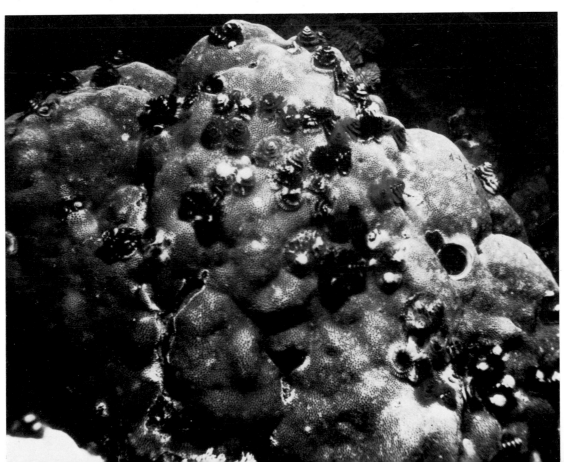

A shoal of black-tipped soldier-fishes (Myripristis murdjans).

3. Fishes

It would be both difficult and presumptious to proceed, within the framework of this book, with a systematic inventory of every species of fish living in the reefs around tropical seas. This incredible fauna comprises several thousands of species and sub-species which would confound many eminent specialists. A complete assessment would necessitate years of work, as well as a collection of some tens of thousands of photographs. Thus, without attempting an exhaustive survey, the principal families of those fishes which are usually met with in living corals will be described, whatever their geographical location. The fauna living in that vast zone, which includes the Red Sea, the Persian Gulf, the Indian Ocean and the South Pacific, is more varied than that found in the tropical regions of the Atlantic.

Each large family of fishes occupies a well-defined place in the reef, so making its identification easier. Pelagic fish, exemplified by sharks, tunnies, barracudas, bat-fishes and sardines, can be distinguished from both non-migratory fishes, such as groupers, lion-fishes and moray eels, and finally, fishes of the coral fringe, such as the little butterfly-fishes.

It is necessary to have some guide lines of identification before diving in tropical seas. In actual fact, many of these fishes are extremely poisonous, so it is advisable to recognise the family, if not the species, to which the animal belongs before touching or eating it. A great many fatal accidents occur each year due to stings (from the stone-fishes, for example), or to poisoning. A short while ago, in Sumatra, a grilled puffer-fish killed an entire family of tourists. With improved transport facilities, which permit numerous beginners to discover the splendours of tropical seas, it is imperative both to make them aware of the dangers and to arouse in them a respect for nature.

Sharks

Sharks are reputed to be dangerous. Inhabiting all the seas in the world, they make up a very numerous order (Pleurotremata) whose individual species are often difficult to identify. Divers often encounter them in tropical seas.

Two main categories can be distinguished: pelagic sharks of the open sea, and lagoon sharks.

The former are fearsomely armed. They have a tremendous capacity for devouring things; it is not uncommon to find in their stomachs the strangest objects, such as bottles of beer or tin cans. These large sharks have the peculiar ability to store part of their food; this enables them to accumulate reserves for their great ocean crossings. The food is then digested in stages, depending on the needs of the organism. If the shark is lucky enough to come across a new prey before it has completely digested its reserve, it will attack, expelling its undigested remains by turning its stomach inside out, and in this way will replenish its 'larder'.

The great white shark (*Carcharodon carcharias*), which measures up to 30–40 ft (10–12 m) in length, is a member of the Isuridae family; it is the direct descendant of the enormous *Carcharodon megalodon* of prehistoric seas, which reputedly measured up to 100 ft (30 m) in length! Fortunately for divers, the great white shark is rare. Peter Gimbel, in his film 'Blue Water' White Death', has alone succeeded in filming it. He sailed round all the great reefs in the world for two years before happening to sight them along the Australian coasts. Their features are straight forward: they can reach a length of 40 ft (12 m), weigh over 5 tons, and charge at up to 45 mph (70 km/h). There is no risk of meeting this animal of the open sea while bathing in a Polynesian lagoon or along the shores of the Indian Ocean.

The blue shark (*Prionace glauca*) is less ferocious than the former. It also is a very large animal. It usually measures 10–13 ft (3–4 m) but can exceptionally reach 20 ft (6 m) in length. Its back is blue, its belly white; it sometimes frequents temperate seas, such as the Atlantic and the Mediterranean. It does not charge, but can start to circle around a person, especially if he is carrying a wounded fish. In this case, it is best not to persist, but quickly to take the fish out of the water.

The small white shark (*Carcharinus albimarginatus*) is easily identified by the white markings on the tips of its fins. Its back is grey, its belly whitish. It can reach 10–15 ft (3–4 m) in length. It is powerfully armed and always hangs around the lower slopes of coral reefs. Young sharks, however, often meet in packs in the shallow waters of lagoons, letting their dorsal fins brush the surface of the water. This creature is a nocturnal predator. Like all sharks, it does not have an air bladder, and is destined to swim all its life in danger of foundering (many sharks, in a state of fatigue, rest on the bottom, sheltering under a coral slab). They form part of the Indo-Pacific fauna, and are in evidence from the Red Sea to Polynesia, by way of Hawaii.

The Pacific black tip shark (*Carcharinus melanopterus*) has a rounded snout. It is a loner, always on the prowl: in Poly-

nesia, when a hunter has harpooned a fish, it is always the first on the scene. It is very timid, and is not reputed to be a man-eater. If, while diving, one of them should appear out of curiosity, the diver only needs to make a loud bubbling noise or to pretend to load his gun to put it to flight.

The grey shark (*Carcharinus menisorrah*), as its name implies, is uniformly grey all over. Although it can measure 10 ft (3 m) it never becomes very big. Always keeping to the outer parts of reefs, like the black tip shark, it lies in wait for a wounded fish. The Polynesians call it 'Mao raira'. In the atolls of Tuamotu, grey sharks move about in packs. When several of them start circling round a diver, he must immediately get back to the boat.

The white tip reef shark (*Triaendon obesus*) mostly frequents the Red Sea, but is also met with in Polynesian waters. It is easily recognised by its rat-like features, its very slender body, and a white patch at the tips of its dorsal and caudal fins. This little shark is very peaceable, and often circles round the diver, but without hostility.

Lagoon sharks are much less dangerous than large pelagic sharks. Their teeth are small compared with those of the deep water sharks. The adults stay in the vicinity of the reefs and are particularly fond of the warm lagoon waters.

The nurse shark is present in all tropical seas, and forms part of the Ginglymostoma family. A nocturnal predator, it is equipped with nictitating eyelids like the Mediterranean dog fishes. Its teeth are not big, but are very sharp, and can inflict serious wounds. This creature, however, is not aggressive. During the day, it is often to be found sleeping beneath a slab of coral. The two main species are the *Ginglymostoma cirratum* in the Red Sea, and the *Ginglymostoma brevicaudatum* in Polynesia.

The shark-like ray (*Rhynchobatos*), also called the guitar fish, lies midway between a shark and a ray in its general

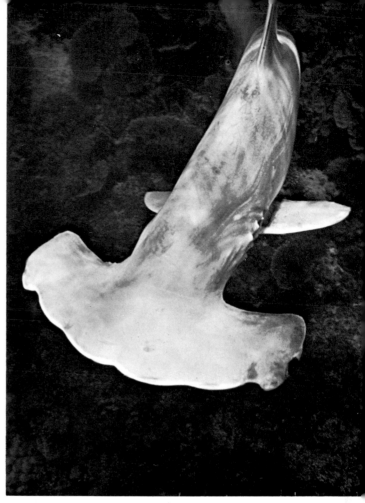

The small white shark (*Carcharinus albimarginatus*) can reach 20 ft. (6 metres) in length. It can be identified by the white patches at the ends of its fins.

The hammerhead shark (*Sphyrna Lewini*) is identified by the characteristic shape of its head.

The Pacific black tip shark (*Carcharinus melanopterus*) is identified by its rounded snout and the black patches at the end of its fins.

Despite its fearsome appearance, the nurse shark (*Ginglymostoma cirratum*) is not dangerous to man.

Guitar fish (*Rhynchobatous djiddensis*) on the lookout for shell fishes.

The Leopard shark is relatively rare. It has no teeth and is quite harmless.

The tiger shark (*Galeocerdo cuvieri*) is a renowned
man-eater.

The harmless white tip reef shark (*Triaenodon obesus*)
is characterised by its rat-like snout and the little
white markings on its fins.

shape and its flattened appearance. It is in fact a sand shark whose dentition is very similar to that of the rays. It is chiefly encountered at night, scouring around the sandy beds looking for shellfishes or crustaceans.

The Carcharinidae family is very large: worthy of note are the mako (*Isurus oxyrhinchus*), sadly renowned along the Caribbean coasts; the *Carcharinus longi anus*, the *Leucas*, as well as the famous tiger sharks (*Galeocerdo cuvieri*) recognisable by the tigrine stripes on their back, and by their obliquely projecting serrated teeth with fluted edges; and finally, the hammer-headed shark (*Sphyrna lewini*), which can reach 16 ft (5 m) in length.

Rays

Two large families of rays live in tropical waters: the Mobulidae, called 'devil-fishes', and the sting rays. The *Manta birostris* belong to the former family. The finest specimens have a span of up to 20 ft (6 m) and weigh up to one ton. They are peaceable creatures, and are happy to eat plankton which they flatten in their huge ever-open jaws with their labial appendages. They can be approached at close quarters without danger, although they are very frightening to watch. During the spawning period, they band together in great shoals of a hundred or more in well-defined areas, where they start cutting strange capers. It is a fantastic sight; sometimes they leap out of the water, probably to rid themselves of parasites, then fall back again in huge columns of foam. They are found in all tropical waters throughout the world. They are often accompanied by pilot fishes or remoras.

There are two types of Manta ray, the enormous *Manta birostris* and the smaller *Manta diabolus*, which mainly move along the South African and Polynesian coasts.

Sting rays are most common in tropical waters. Their tails are dangerously furnished with a fluted spine, which they use as a whip with incredible precision. Sting rays, like all flat fishes, instinctively settle in the sand to camouflage themselves. Therefore, it is advisable never to set foot on a sandy bed without having inspected it well beforehand.

There are many kinds of sting rays, but however much they differ, they all have a spine positioned under the tail. The most common are the coach-whip ray (*Himantura*), the spotted or eagle ray, identifiable by its very long tail flecked with white spots (*Aetobatis nari nari* and *Myliobatis*), the *Taenuria* of the Indo-Pacific Ocean, grey with splendid little blue dots, and the *Taenuria lymma*, whose flesh is succulent.

Rays are peaceable creatures which only use their spines when they think that they are being attacked. Despite this, their sting can be fatal if delivered to the stomach: a potent injection of venom is always coupled with the wound from a spine. The underwater fisherman would be ill-advised to harpoon a very large ray: these creatures are imbued with considerable energy and such powers of resistance that the attempt would be doomed to failure.

This enormous ray (*Manta birostris*) is several feet wide and weighs a ton. It is accompanied by three remoras clinging to its belly.

Grouper, known as the 'carpet cod' in Polynesia, lies
hidden between two slabs of *Acropora* coral.

A shoal of hatchet-fishes (*Parapriacanthus*) at the entrance to a grotto.

The *Gaterin niger* has gilt patches over its skin which appear or disappear according to its mood.

These sweet-lips (*Gaterin gaterinus*) form a solid unit so as to feel secure.

Hatchet-fishes

They are diurnal dwellers in the dark grottos of the reef and always live in very compact shoals. When danger is imminent, they emit a sort of cracking sound. Their 'V' shaped body has earned them the nickname hatchet-fishes. They are planktivorous and only come out of their shelter at night to feed.

The *Parapriacanthus guentheri* are little silvery transparent fishes, living in very compact shoals in front of the entrance of grottos. They are often found accompanied by black groupers which hide behind their protective screen. Jacks often burst in, snapping up some of the fishes as they pass through the shoal. These hatchet-fishes are found around branches of brown gorgonians, called black coral by the natives of the Red Sea.

Bat-fishes

These are often found under large wrecks in tropical seas. The young fish clearly differ from the adults both in their shape and their behaviour. In fact, at birth they have a large membranous growth on their fins. They are not very timid, and like the still harbour waters, being particularly fond of the cables of casting anchors. When they become adult, their fins contract, and they begin to lead a pelagic life, often in very large shoals.

They are in evidence in every tropical sea in the world. They are sociable creatures, and seem especially fond of the company of divers. Their flesh is grey and completely inedible. On the other hand, they are in great demand as aquarium species, because they can live for a long time in captivity. Besides, they are very easy to tame. There are many species of them.

The most common species, the *Platax pinnatus*, is found from the Red Sea to Polynesia. When young, its large fins are crossed by black vertical stripes. Later on, it looks like a large grey butterfly-fish (*chaetodon*). It is in regular attendance around shipwrecks and can measure up to 25 in (60 cm) and weigh 30 lb (15 kg).

The *Platax orbicularis* prefers to live among the coral formations. It never becomes very big and its appearance is generally more pleasant than that of its cousin. It is characterised by more streamlined fins and a black vertical line located three-quarters of the way up its body.

Sweet-lips

These are also called gaterins, and are usually found in all tropical seas. Like the snappers, they assemble in very compact shoals behind clumps of corals, sheltering from the currents. They sleep like this, eating nothing all the day through. At night, sweet-lips separate from the shoal to feed. This enables the diver to observe the dispersal of these fantastic gatherings at twilight.

They are carnivorous fishes, which forage fairly deep down on the sandy beds. In the morning, they come back in order to re-assemble into a compact shoal.

Concentrations of sweet-lips are common in the Red Sea and the Indian Ocean. In Polynesia, they are called 'ataras'. Their flesh tastes insipid. Besides, they look so beautiful in the water and seem so peaceable that it would be a real pity to hunt them.

There are three main species: a large grey one (*Gaterin niger*); a yellow one dotted with black spots (*Gaterin gaterinus* or *Plectorhynchus*), which assembles in the strait of Bab-el-Mandeb in such dense concentrations that they give the impression of forming barriers; they are the most familiar of the sweet-lips family and are often found in very warm seas, the Red Sea in particular; and finally the *Gaterin orientalis*, which has a yellow head and is streaked with black horizontal bands over its white skin. This species is very common along the great reefs of the Indian Ocean (the Comoro Islands, Madagascar). As it is the most timid, it is the most difficult to observe.

A shoal of bat-fishes (*Platax pinnatus*).

Top left: A pair of butterfly fishes or chaetodons in the Gulf of Tadjura.

Top right: *Chaetodon melanotus* wandering in the depths of a petrified forest.

Middle left: *Chaetodon benetti* with the strange eye-shaped patch on the back of its skin.

Middle right: *Chaetodon meyeri* 'hovering' over coral clumps.

The lemon butterfly-fish (*Chaetodon semilarvatus*) needs no explanation to justify its name.

This brightly coloured *Chaetodon falcula* seems little impressed by its wild and desolate surroundings. ▶

Angel-fish (*Pomacanthus semicirculatus*): many fine specimens are to be found on the coasts of Madagascar.

Angel-fish (*Pomacanthus filamentosus*) wearing a worried look like a theatrical mask.

Right-hand page: This angel-fish (*Pomacanthus paru*) is one of the most beautiful fishes in the Caribbean.

Angel-fish (*Pomacanthus arcuatus*) living in the Caribbean Sea.

Angel-fish (*Pygoplytes diacantus*): one of the finest reef fishes.

Damsel-fishes and clown-fishes

Damsel-fishes (*Chromis*) and clown-fishes (*Amphiprions*), together with 'barbers', comprise the mini-fauna living above the coral branches. At the least danger, this little group rushes into the coral formation and hides in its deepest parts. If the coral is taken out of the water, the little damsel-fishes stay jammed in it. It suffices only to shake the coral to remove them.

Their main feature is a single nostril on each side of the snout, in place of two. Their skin can be almost any colour, depending on the species.

Four main genera of the Pomacentridae family can be distinguished according to their eating habits: sergeant-majors (*Abudefdufs*), which are herbivorous; planktivorous damsel-fishes (*Chromis dascyllus*); omnivorous damsel-fishes (*Pomacentrus*); and carnivorous clown-fishes (*Amphiprions*), which live in symbiosis with sea-anemones.

When the time comes to spawn, pairs of damsel-fishes choose a piece of ground and thoroughly clean the surface. The female lays her eggs there, while the male vigilantly watches over them until they hatch. The hatched eggs float up to the surface, where the larva leads a planktonic existence for some days. It will then attach itself to a colony of 'acropora', near which it will spend the rest of its life. Some young damsel-fishes, enjoying the same immunity as clown-fishes, succeed in colonising a sea-anemone without being troubled by its stings. When they reach a certain size, they part company and return to dwell between the coral branches.

Six-banded sergeant-majors (*Abudefduf sexfaciatus*), with black and white stripes, always move about the corals in large groups. They shelter beneath them at the slightest alert.

Blue damsel-fishes (*Chromis caeruleus*) always swarm around branches of 'acropora'. They are retrieved for aquariums by taking the coral out of the water and shaking it strongly.

The *Chromis ternatensis* resemble Mediterranean damsel-fishes, but their skin is slightly lighter in colour. They are often found in very large shoals under branches of 'acropora', in which they shelter in times of danger.

Clown-fishes (*Amphiprions*) are the most colourful of all. They are mainly characterised by the fact that they live in symbiosis with sea-anemones. It is not entirely clear whether it is in fact real or antagonistic symbiosis. It is known that sea-anemones live equally well without clown-fishes. On the other hand, the fishes are completely dependant on the *Actinia* to shield their over-conspicuous bodies. Before getting to know a sea-anemone, the clown-fish 'vaccinates' itself against its stings by gradually touching it for the ultimate purpose of smearing its body with a protective mucus which prevents the discharge of stinging cells from the sea-anemone. An experiment carried out by the author clearly demonstrates this fact. Having put the fish to sleep using the drug MS 222, he wiped it all over to remove the sticky mucus covering its scales. Once the animal awoke, he put it back into the sea-anemone, placing it on the tentacles. Despite its desperate efforts to save itself, the little clown-fish was instantly devoured by its own companion. This leads the author to believe that the relationship is one of antagonistic symbiosis. In support of this hypothesis, there also remains the fact that sea-anemones do not rely for their nourishment on the scraps from the meals of their clown-fishes. On the contrary, amphiprions have often been seen to feed on the liquid expelled from the mouth of the sea-anemone.

The two-banded clown-fish (*Amphiprion bicintus*) is found in different types of sea-anemones, but more especially those of the Stoichaetis genus.

The single-banded clown-fish (*Amphiprion ephippium*) is orange with a single blue band on the front of its body.

At night, this two-banded clown-fish (*Amphiprion bicinctus*) shelters deep inside the anenome.

Single-banded clown-fish (*Amphiprion ephippium*) patrolling in front of its anenome.

A shoal of sergeant-majors (*Abudefduf sexfaciatus*) moving about the porite corals.

The two-banded clown-fish (*Amphiprion bicinctus*) is very common in the Red Sea.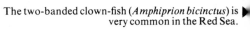

Damsel fishes (*Chromis caeruleus*) playing among the coral branches, ready to hide themselves if the slightest danger threatens.

Damsel-fishes (*Chromis ternatensis*) creating a dense cloud as they travel around the corals.

Gudgeons

Gudgeons are closely related to gobies. They are little sedentary fishes which have only become known through underwater exploration. In fact, they always live on the sea bed, near to a deep burrow that they themselves have dug and where they come to shelter at the least danger. There are often found at the bottom of reef slopes, by the side of a lagoon bordering on the detritus of dead corals and sand. It is certain that they feed on plankton.

The *Nemateleotris magnificus* is unusual because of the way that its dorsal fin spreads out in the shape of a trigger.

The tipped gudgeons (*Ptereleotris tricolor*) live in open water 10–15 ft (3–5 m) up from the sea beds, where they establish their burrows.

The parapercidae

The Parapercidae have recently been connected with the Mugilidae family.

The *Parapercis hexophthalma* is a curious fish: posted at the bottom of the reef, it incessantly follows the diver, rolling its eyes in a comical way when watching him. It should not be mistaken for the lizard-fish (*Saurida*), which takes up the same position on the bottom when on the look-out for its prey, before leaping out, as quick as a flash. It is met with all over the Indo-Pacific region, from the Red Sea to Polynesia.

Parrot-fishes

They are so-called because of both their bright colours, and their teeth which are fused together in the shape of a beak. Contrary to what was long believed, it has just been discovered that these creatures are herbivorous. In fact, they are often to be seen scraping at corals with their powerful beaks. An in-depth study of this phenomenon has shown that they mainly eat microscopic algae growing on dead corals. Nevertheless, it cannot be categorically stated that they do not eat corals, as they have frequently been known to rasp insistently porite polyps.

They are all diurnally active, and sleep soundly at night. During the day, they are very busy animals, swimming from one clump to another, scraping everywhere, not even stopping to defecate. It is impossible to get near enough to them to take good photographs unless they are surprised in the act of eating behind a coral cluster, for they are very timid. At nightfall, they look for a hole in which to sleep. However, while staying faithful to their particular reef, they rarely occupy the same hole twice. During the night, in the Pacific for instance, parrot-fishes cover themselves in a veritable cocoon of venomous mucus in order to protect themselves from predators; but in the Red Sea, they often sleep right on the bottom without the slightest protective device. This bizarre phenomenon has not yet been explained.

There are a multitude of species of parrot-fishes, which are excessively difficult to identify owing to their similarity. There is even disagreement on their names.

In general, parrot-fishes are timid creatures. It is inadvisable to catch them, as their flesh is often poisonous. It is best to let them live in peace. Besides, they exercise a significant ecological function by contributing to the manufacture of coralline sand.

Top left: Parrot fish (*Scarus sp.*) which owes its name both to the vivid colour of its skin and to the shape of its beak.

Top right: Australian *Scarus*, coloured blue and growing to a large size.

Middle left: Scratches on a porite coral left by the beak of a parrot-fish.

Middle right: Gudgeon (*Nemateleotris magnificus*) having the shape of a trigger.

Bottom left: Gudgeon (*Ptereleotris tricolor*) considered for a long time as a very rare fish.

Bottom right: The *Parapercis hexophthalma* is a merry little fish, bursting with curiosity.

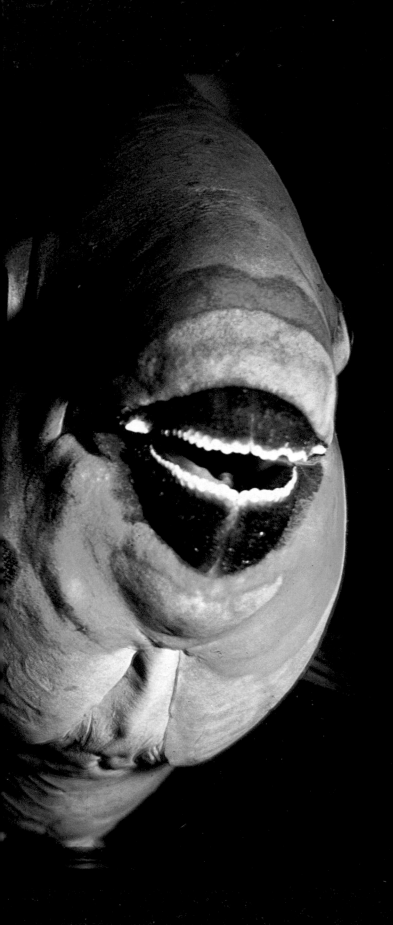

Parrot-fishes.

Turtle (*Chelonia mydas*) at rest in the coralliferous deeps.

4. Marine reptiles

Turtles (or chelonians) constitute an order of the class of reptiles. They lead an entirely aquatic life, except when they lay their eggs. When this time comes, turtles come out of the water and dig a hole on the shore, using their hind legs. They lay several hundreds of eggs, which they then cover with sand. But this ordeal exhausts them, since their weight increases enormously out of water, and many die.

As soon as they are born, the little turtles go in the other direction towards the sea. They swim instinctively, but only gradually learn to dive, and it takes several months before they reach even moderate depths.

Of the three living species, the hawksbill turtle (*Eretmochelys imbricata*) is highly prized for its thick plates, which are used for making jewellery. Furthermore, they are reared in large tanks from eggs which have been gathered from the shores. Up to 10 lb (5 kg) of plates can be obtained from the largest specimens. This turtle is carnivorous, living on crustaceans, sea-anemones and various molluscs, which it breaks open with its horned beak.

The green turtle (*Chelonia mydas*) resembles the hawksbill turtle, except that its plates are less thick and are placed side-by-side, not overlapping. It is hunted for its meat.

The leathery turtle (*Dermochelys coriacea*) has a very thick hide in place of a carapace. It is a real monster and can span 10 ft (3 m) and weigh 1100 lb (500 kg). It is solely a pelagic species and is rarely seen near the coast.

Hawksbill turtle (*Eretmochelys imbricata*).

A little clown-fish valiantly defends its anenome.

It charges at the attacker.

A turtle diving.

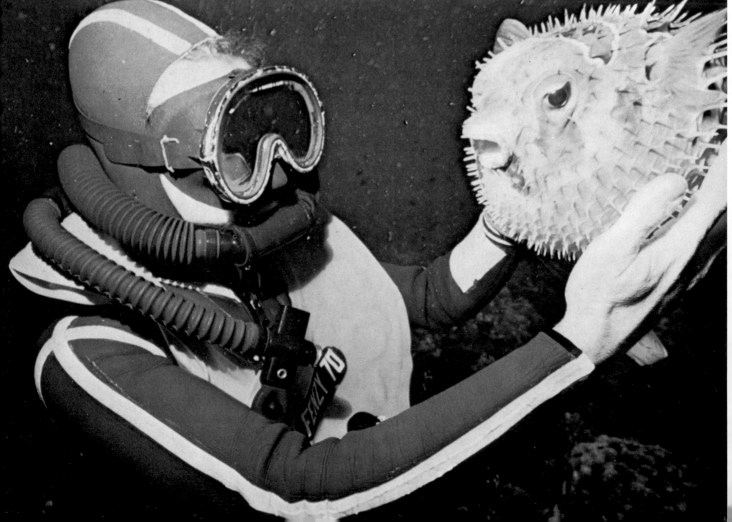

Underwater photography

A perfect mastery of diving techniques is essential for anyone who wants to become a good underwater photographer. Completely proficient in his diving techniques, he will be able to concentrate his utmost attention on both operating his camera and getting close to animals. At night, the diver will find it easier to catch sea creatures unawares. Some, waking with a start, will remain transfixed in the strange light; others, intercepted while in full pursuit, will flee in search of a shelter from where they will be able to watch the intruder, looking either sullen and suspicious, or else puzzled and perplexed. Each watches the other ... curiosity often carries the day, but contacts are broken with unforeseen abruptness. The image-chaser must therefore know how to take advantage of this immobility at every opportunity, as the speed of a fleeing fish can be too fast to be 'caught' by the camera. Fishes have the remarkable ability of disappearing on the spot.

The underwater photographer should be sufficiently proficient to be capable of diving safely on his own—but then only with a compressed-air inflated life jacket. Use of an aqualung gives him more time to concentrate on his photography than if he is snorkelling. Even so, no-one should ever dive deep or at night without a companion. For night diving, a powerful 100 watt lamp fitted with a special 6 or 12 volt Ni-Cad battery dazzles the animals. For diving during the day, at whatever depth one works, it is absolutely necessary to use flash equipment to bring out the warm colours, which disappear in natural light in the first few feet. There are two ways of using flash lighting: it can either be fixed directly onto the camera or be carried by another diver. To avoid the flash illuminating particles floating in the water it should be offset from the central axis of the lens, either to one side or above. Well-balanced lighting, accurately applied (at an angle of 45° to the optical axis) will contribute ninety per cent to a successful exposure.

The taking of underwater pictures is fraught with three major problems, all due to optical phenomena peculiar to aquatic surroundings: the lack of light and contrast, lack of depth of field, and disappearance of colours. Diffusion and absorption of light are responsible for hazy results, a monochrome picture and a 'flat' result. In general, the exposure should be 1/60th of a second for a fixed subject, and 1/125th of a second for a moving subject. In fact, actual conditions are infinitely variable and depend on the equipment used. When using flash the apperture should be opened by two, three or more stops than the guide number calculator on the camera states. Close-up shots are best, with the subjects 3–10 ft (1–3 m) away. A subject in the water appears to be bigger and nearer the camera than it actually is. Thus, a subject which in reality is 4 ft (1·33 m) away appears underwater to be only 3 ft (1 m) away, the refractive index of water being 4/3 or 1·33. For cameras with adjustable focus, the distance to set it to will be the apparent distance underwater 3 ft (1 m), rather than the real distance 4 ft (1·33 m). Water rapidly absorbs colours, depending on depth. Absorption begins with red; the last colour to disappear is yellow. In practice, depending on the water—coastal or ocean, tropical or temperate—red begins to disappear almost immediately. Unless artificial lighting is introduced, within 20 or 30 ft (7 to 10 m), all colour pictures will have a greenish-grey cast. In practice, there will be a diminution of red, which will completely disappear at 16 ft (5 m), and orange, at 30 ft (10 m).

Using a camera to take macrophotographs.

180

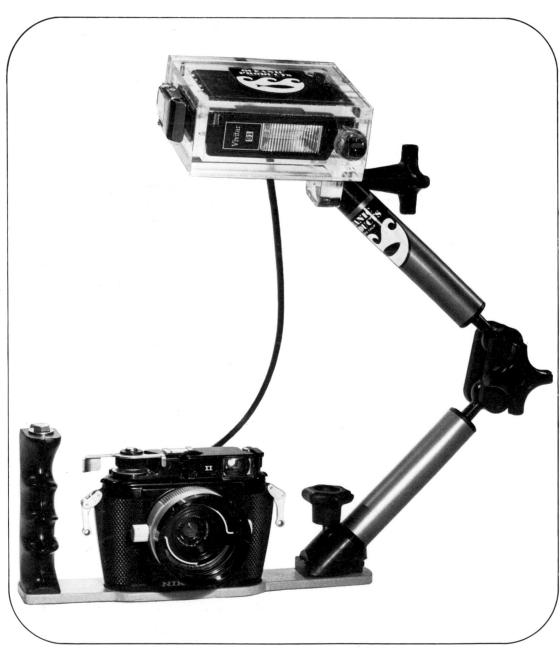

A Calypso Nikkor II camera equipped with the
Oceanic Product outfit. The complete system is
relatively economical and can be fitted with almost all
makes of flash equipment.

I. Photographic equipment

Films

In clear, well-lit water

KODAK PLUS-X or VERICHROME -PAN is best suited for black and white underwater photography at shallow depths without using lighting. If the depth increases and the right type of camera is at hand, it is best to use a KODAK TRI-X 400 ASA film, or even a KODAK RECORDING 2475 film of 800 ASA. With colour, fine prints can be obtained by taking pictures on a KODACOLOUR II 80 ASA film, without going any deeper than a few feet (metres). It should not be forgotten that the colours of subjects tend to fade, as water filters light, eliminating first reds and then yellows, leaving only blues. KODACOLOR film has the advantage of being available in all sizes. Transparencies or cine-films taken beneath the surface with KODACHROME Daylight—25 ASA (PHOTO) or 40 ASA (CINE SUPER 8) film corrected to 25 ASA by the filter on the camera, will be very beautiful.

At a few feet (metres) down, KODACHROME 64 ASA or KODAK EKTACHROME 64 ASA films allow more freedom to photograph. Finally, at even greater depths, it will be profitable to use KODAK EKTACHROME HS, Daylight Type 160 ASA film for photographs, and KODAK EKTACHROME 160 CINE film at the same speed for cine-films. These films allow exposures to be made in remarkably unfavourable lighting conditions.

Cameras

There are two methods of taking underwater pictures: with a watertight camera; with a normal camera, placed in a watertight housing.

The watertight camera

There is one principal watertight camera: the Calypso Nikkor II: this extraordinary apparatus is very small and compact, and can be used equally well on land as underwater. It is sold with a 35 mm lens which is reduced to a focal length of nearly 50 mm underwater because of refraction. Three other lenses can be fitted to it: a remarkable super wide-angle 15 mm lens, a 28 mm wide-angle lens, and an 80 mm telephoto lens.

Land cameras in watertight housings

It is advisable, when first starting to take underwater pictures, to use a small uncomplicated good quality camera. The results will be satisfactory down to a depth of 15–25 ft (5–8 m) in clear, sunny conditions. Beyond this depth, it is preferable to use flash equipment if both the camera and the housing have been designed for it.

Similar housings are available commercially in other countries.

Lenses

A normal surface lens behaves like a telephoto lens underwater if it is used behind a flat porthole. Then its focal length should be multiplied by 1.33; thus, a normal 50 mm surface lens becomes an 80 mm telephoto lens under water. For panoramic shots, a wide-angle lens should be used.

Under water, surface lenses become as follows when used behind a flat port:

Camera	Wide angle	Standard	Tele-photo
6 × 6 cm	40 → 50	80 → 100	150 → 200
24 × 36 mm	35 → 50	50 → 80	100 → 150

183

The most significant discovery ever made in underwater photography is that a cheap plastic dome can be used as a porthole and that it will correct the effects of refraction on the angle of view of a lens. This has resulted in a dramatic improvement in quality, especially in the poorer visibility of British waters. Indeed, the improvement is so marked that now it is pointless to buy any underwater housing unless it is capable of being fitted with a dome porthole—unless one proposes to use the outfit only for extreme close-ups or with a telephoto lens.

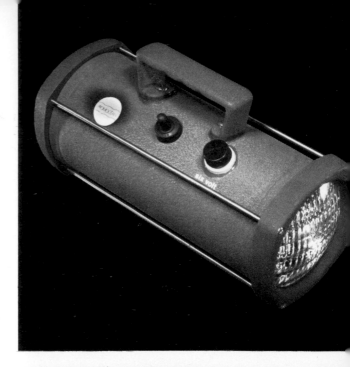

Promocean housing for a Kodak Instamatic XL33.

Bottom left: Promocean housing for most Kodak cameras.

Bottom right: Promocean housing for a Kodak Pocket Instamatic.

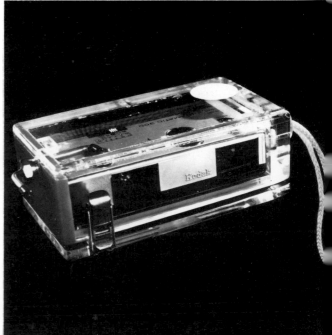

◄Promocean light source for
underwater use.

Oceanic Products underwater
outfit designed for use with a Nikon
F2. This remarkable outfit is fitted
with an electronic eye to regulate
the flash exposure.

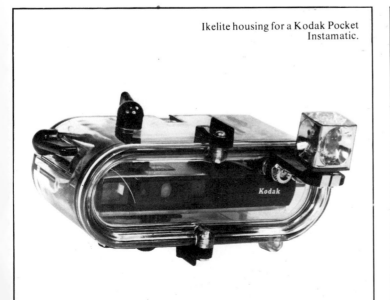

Ikelite housing for a Kodak Pocket
Instamatic.

Ikelite housing for XL33 and XL55
cameras.

Diving in an aqualung.

2. Photographic techniques

Macrophotography

Macrophotography consists of repro-ducing very small subjects larger than life. It gives particularly interesting results when depicting the underwater world, as it shows the large variety of shapes and colours in the smallest detail. To achieve these shots, focus must be extremely accurate, the stability of the photo-grapher being essential: he must be equipped with a weight belt which will help him to maintain negative buoyancy. He should look for a support to rest on and should bring his camera to bear on the chosen subject without disturbing the sea bed.

To obtain the maximum depth of field it is essential to use a small apperture—at least f16—and this also requires the use of flash except on rare occasions. The easiest way to compose the picture is by means of a focusing frame fixed to the front of the camera or housing in such a way that it indicates the plane of sharp focus and also the picture area.

On every modern photographic apparatus with interchangeable lenses, it is possible to interpose macrophoto-graphic extension tubes between the lens and the camera in order to modify the focal distance. Close-up lenses or sup-plementary lenses can also be used, par-ticularly on the CALYPSO NIKKOR; these are tricky to use on lenses with long focal lengths. Extension tubes give an enlarged image size with an ordinary 35 mm lens.

Cine

At the present time, there is no watertight cine-camera on the market. Watertight housing must therefore be used.

The list below gives KODAK cameras with suitable watertight housings:

Super 8 Kodak Cameras	Housing
Instamatic M 22	Fulgor Mare
Instamatic M 24	Promocean Sea 3
Instamatic M 26	
XL33 Movie camera	Promocean Sea 4
XL55 Movie Camera	Imasub-Submatic

Lighting is an important consid-eration. It must provide constant illumi-nation, which involves the use of a 100 watt quartz iodine lamp.

The stability of the cameraman is as important, if not more so, as it is when taking still shots. The housing should be capable of being held firmly in the hand.

187

Nudibranch (*Glossodoris*).

3. Successful photography

The attainment of excellent pictures of underwater creatures will only come with the experience acquired from numerous observatal dives and after numerous set-backs. The main problem consists in getting as near to the fish as possible, so as to reduce the effect of diffusion, whilst trying to avoid putting it to flight.

The methods employed vary from species to species, and can at times unexpectedly involve a companion. Generally speaking, it is important to find a distraction which diverts the attention of the animal and excites its curiosity, so that it forgets the presence of the diver.

Wrasses and small serranids
A single broken sea-urchin will entice these alert and voracious creatures, which are prepared to subject themselves to tremendous dangers for the sake of a few scraps. The diver is then able to get very close to them, but must act promptly, since as soon as they have quickly gobbled up their prey, wrasses and small serranids rush away, vanishing into a crevice.

Octopuses
They pose with great docility. Their powers of mimicry can give rise to a series of amazing photographs if they are caught lying on different beds. The diver can stage a fight between an octopus and a moray eel, using the battle as a pretext for excellent documentary pictures.

Crustaceans
Macrophotographs of the head of a crustacean always produce worthwhile results. The peduncules of a crab can create a vivid impression on the novice. Supplied with tit-bits, the diver can coax his models out. Crawfishes are particularly timid, so it is difficult to get closer than 15 in (40 cm) to them. To lure one of them out, the diver can offer it a dead fish, but it must be given time to tell that it is food by sensing it with its feelers. A prey seized in the formidable claws of a crab can present the opportunity for a good picture. If he is lucky, the diver might come across a hermit crab carrying a cluster of sea-anemones on its shell; if, frightened by the presence of the diver, it retreats into its shell, there will be time to get an angle on it and wait until it re-emerges.

Sea breams and saupes
The best way of catching these fishes unawares is for the diver to stage a 'feint'; that is, to noiselessly sink down into the water holding his breath, and to crouch between two rocks, keeping perfectly still. The fishes, feeling reassured, will come and pass right in front of the lens.

Groupers
It is difficult to describe the behaviour of a grouper towards man. This carnivorous fish is often fearlessly inquisitive, accompanying and following the diver for long periods, and watching his slightest movements. It is unperturbed by the noise made by a shutter release or the crackle of a flash gun. It will even come and take a fish out of the diver's hand.

Congers and moray eels
Taking close-up shots of the heads of conger and moray eels poses no special

problems; these animals stay in a fixed position all day long, with their heads sticking out of the hollows in which they live. At night, their manoeuvres can be photographed without trouble. It is more difficult during the day, since they return to their holes; nonetheless, with a little patience, the diver can induce them to come out by waving a morsel of fish in front of their snouts. With large tropical moray eels, he should take care for his hands: these fishes seize their prey with incredible ferocity, and once their retractile fangs are entrenched, they do not easily let go of their catch.

Little sand sharks

There is an excellent way of taking dramatic photographs of sharks. Little nurse sharks sleep soundly during the day, embedded in a hole, only letting their long tails show out. Another diver should be asked to pull one of their tails, and as soon as the animal is released, to grasp the top of its head while supporting it with the other hand. In this way, a magnificent photograph can safely be taken of the 'fight' between a man and a shark but it would be wise to choose a small shark.

Large sharks

The photography of large sharks has always posed many problems: the main disadvantage lies in the fact that these creatures do not allow anyone to come closer than 12 or 15 ft (4 or 5 m) to them, so photographs of them are always shrouded in blue, which reduces the contrast. There is only one solution: that is to find a way of drawing near to the subject. It should be made clear that, before setting out to photograph sharks, it is necessary to know the places where they are sure to be sighted. This is not as obvious as it seems. Some oceans, like the Pacific, are particularly rich in sharks, and finding them is no problem, whilst others, like the Red Sea for example, only harbour sharks seasonally. Yemeni fishermen know this well. Therefore, before preparing an expedition to photo-

graph sharks, it is necessary to make enquiries from the local hunters. There are two methods of approaching sharks. The first, which is suitable for sharks of the open sea, necessitates the use of a cage, a winch, and a large boat. The diver arranges for a massacre to take place with several large fishes, and witnesses the rush for the spoils from his submerged cage. This can provide the pretext for taking some very good pictures. The second method, used for reef sharks, involves being completely familiar with diving. The diver, assisted by two well-armed divers, will entice the little sharks with a bait of wounded jacks. He should be very wary and alert, as in the ensuing frenzy an accident is easy to occur, since the charge of a shark has the speed of lightning. He alone should judge whether it is reasonable to continue the operation or whether it is preferable to climb back on board the boat. It is obvious that only an experienced diver can carry out these types of manoeuvres; there should always be a good look-out crew in attendance at the surface. *But always remember: no shark is safe—even a small one!*

Scorpion-fishes and stone-fishes

These creatures have a habit of lying motionless on the bottom and can be extremely difficult to see. They lie in wait to swoop on anything that passes within range of their mouths. They will always be found lying in wait in this position. No great precautions will be needed to approach them. It is better to take a close-up of the head, which, with its fleshy excrescences, makes them look just like dragons. The diver should be careful to avoid touching either fish as the venom of their spines is so toxic that it can cause poisoning, which can even be fatal in the case of the stone fish.

Shoals of fishes

Photographs of shoals of fishes are certainly the most difficult to take, but they are by far the most beautiful. A camera with a wide-angle lens should be used. The principal problem is to succeed in getting sufficiently close to, and even to manage to penetrate the shoal itself. To obtain a clear picture, the distance between the lens and the subject must be reduced to the minimum so as to avoid diffusion of artificial light by particles suspended in the water. It is always tricky getting into the midst of a mass formation of fishes. There is no general rule, as the method varies according to the species. However, some guidelines are valid for all fishes. Once the shoal is located, it must be approached slowly and without any brusque movements; several 'feints' might be necessary; if there are no sharks about, some morsels of fish will entice it (but there is a risk of disturbing the water). Above all, much patience is needed and the diver should be prepared to stay for half-an-hour without moving. The way that they react to danger differs according to the fishes themselves: some tightly close up their ranks when danger threatens, so as to form densely compact spheres; others immediately disperse. With the latter, the camera should be carefully set, for at the first flash, they will flee, making a rustling sound with their scales, like a flight of pigeons.

Corals, gorgonians, sea-anemones and echinoderms

If approaching fishes seems too risky, very fine shots can be taken of specimens of fixed fauna, such as coral polyps and species of *spirographes*. Yet, it is still important to operate slowly, as these creatures retract at the slightest sudden movement. The fantastic wealth of colours and shapes of corals is fascinating. Tropical sea-anemones with their symbiotic associates, the clown-fishes (*Amphiprions*), are always entrancing to photograph. Should the diver suddenly draw near to the sea-anemone, clown-fishes, reacting defensively, will rush forward to meet him, butting against him with their tiny snouts, whilst emitting harsh sounds in the water. When they do

not receive any reaction from the diver, they panic, and rush to shelter among the tentacles of their companion where, tucked away like babies, they allow themselves to be photographed. But care should be taken if touching them as some tropical corals can sting.

Teaching underwater photography

An international centre for underwater photography has opened on the Mediterranean coast. Fitted out like a real diving resort (with boats, compressors, possible accommodation), it possesses in addition a very considerable colour laboratory, which processes films within a day and cine-films in 24 hours. Each member can also inspect the results of his labours as his dives proceed, and carry out numerous tests, using his own equipment. All through the year, fairly intensive courses are held under the technical direction of the author. In America, there is an exceptionally well-equipped school of underwater photography, Seacor Incorporated, managed by Jim and Cathy Church. At the end of the course, it issues a certificate giving access to the Underwater Photographers Club of America. Many underwater safaris are organised in specific areas of the tropics. The actual location is chosen for its specialised interest: abundance and variety of animal life, migrations of fishes, spawning grounds, and so on. A small team of participants can produce very good animal documentaries at the most favourable time of the year with the best technical resources at their disposal (boat, compressor, generating set, projector, cine-camera etc.). Every year in April, a special safari is organised at Djibuti. It is called 'Shoals of fishes in the Farasan Islands', and coincides with the migration of jacks. In the Comoro Islands, during mid-September, a safari takes place at the time when turtles hatch on the Moheli Island. In the Maldive Islands, expeditions devoted to the discovery of underwater grottos and coral atolls, are undertaken from October to April.

The joys of diving.

Diving

The conquest of the underwater beds presents numerous problems for man. He must adapt himself to an environment for which he was not made. Pressure underwater is much stronger than atmospheric pressure. The conditions for the assimilation of gases by the human organism are completely altered.

Equipment

Cylinders of compressed air fitted with a demand valve provide the diver with air at a slightly higher pressure than that of his surroundings. It increases by one atmosphere every 33 ft (10 m). There are demand valve regulators in one or two stages, meaning that the reduction of the high pressure of the air in the cylinder to the consumption pressure can be made in one or two steps. An aqualung is generally fitted with one, two or three cylinders, each with an average capacity of 70 ft³ (2 m³). For instance, one can spend half an hour at a depth of 100 ft (30 m) without having to go up to the surface in stages; now, 70 ft³ (2 m³) is not usually enough air to spend more than half an hour at 100 ft (30 m) down, except for an experienced diver with perfect control over his breathing. A beginner should therefore be satisfied with 70 ft³ (2 m³) and once he has assimilated all the techniques of diving, he should be able to descend with more air for a longer dive duration.

Diving cylinders are made of steel. They are fitted with various taps and valves, including a reserve appliance consisting of a spring set at 65 lb (30 kg) which, when the residual pressure goes below this limit, shuts off the air access. The diver then experiences difficulty in breathing, and once he has opened the valve, is compelled to surface. Steel cylinders are heavy, but light alloys have for a long time been banned after explosions occurred. Nevertheless, they are again being made in the United States and Germany. In France, cylinders must be submitted for inspection by the Mining Department every five years.

To photograph under the best conditions, the diver must be equipped with an inflatable jacket which enables him to steady himself at all times. His mask should be carefully chosen: a badly fitting mask can become unbearable after two or three hours through pressure on the forehead and on the top lip. A rubber diving suit can be useful to the diver, but it is not essential. It enables a constant temperature to be perfectly maintained even after two or three hours of immersion. Besides, it effectively protects the diver against all manner of cuts and stings. Lastly it can give a feeling of security when face to face with sharks . . . However, the diver in tropical waters need only wear a pair of trousers with braces, a linen shirt or a vest covering his chest and back. A pair of flippers and a snorkel (necessary for swimming on the surface) will add the finishing touches to his outfit. A knife is an important safety device, especially for the lone diver. He can also use it as a tool for digging or lifting up rocks. Gloves are not superfluous, as they protect the hands from a number of painful cutaneous accidents.

The dangers

Good equipment alone is not enough to overcome all obstacles and avoid every danger. The ways in which the environmental laws of the sea affect human physiology should be briefly recounted.

A novice diver feels a sharp pain in his ears as soon as he goes below a depth of 5 to 10 ft (1.50 to 3 m). This unpleasant sensation is caused by the ear drums being distorted by the pressure of the water. In order to restore the equilibrium between the internal and the external ear, he should squeeze his nose and try to blow out at the same time (this is the so-called 'vasalva' method). When he becomes more used to diving, a simple swallow will suffice. Gases are not diffused in the same way in water as they are in air. Nitrogen

197

in fact tends to go more into solution in the diver's blood as he goes deeper and the pressure gradually increases. When he rises to the surface, the nitrogen comes out of solution and is given off in the form of gaseous bubbles in the blood stream if decompression is too rapid. Therefore, the ascent to the surface must take place very slowly, and the duration of pauses, stages of decompression, should be gauged according to the depth and the time spent at that depth. There are tables on sale that indicate the duration of pauses according to the length of time and the depth of the dive. It is very useful to have a decompressimeter, that automatically indicates the stages of ascent that must take place.

The symptoms of decompression sickness are as follows: itching, a rash, a tingling sensation in the joints, numbness, fatigue, sickness, paralysis or dizziness. If the diver is only slightly affected, it will suffice to re-immerse him, so that he can re-surface after going through the requisite stages. If he is seriously afflicted, the use of a decompression chamber and his immediate transfer to a specialised hospital centre should be implemented. When diving at great depths, a man can show signs of drunkenness, owing to the accummulation of toxic gases. As soon as the first symptoms manifest themselves (difficulty in controlling his reflexes and movements, an abnormal feeling of euphoria), he should surface slowly. This phenomenon usually results below 200 ft (60 m).

The stroller underwater does not run any more risks than his terrestrial counterpart. Nevertheless, a good knowledge of the animal life will help him to avoid any unpleasant occurrences, if not serious accidents. Most injuries are caused through negligence and clumsiness. Many aquatic animals are armed with venomous devices ready to strike at the least provocation, particularly the *Actinia*, jelly-fishes and corals. The diver who hunts, captures or conveys fishes runs the risk of being poisoned by spines,

fins and opercula: he should always handle them carefully. The most dangerous species are as follows: weeves, scorpion-fishes and fire-fishes, sting rays. Inherent aggression is a controversial subject. Some sharks can, however, be formidable adversaries. It is therefore best to act very cautiously in waters where they often abound. The crew or the passengers on the accompanying boat should be always on the alert, ready to receive and quickly hoist up the diver who finds himself in difficulties. A profound knowledge of the species which can be encountered in diving sites will enable him to familiarise himself with the patterns of behaviour and the reactions of each of them. When their attitude seems to be alarming or dubious, the prudent diver will beat a retreat and leave the water.

Diving sites

Diving conditions differ according to the latitude. Man's repeated aggressions endanger the equilibrium of the marine world. Both fauna and flora are becoming progressively scarcer along European coasts. In the Mediterranean, it is only at comparatively great depths that underwater life continues to expand: a dive is only worthwhile if it is made from a depth of at least 130 ft (40 m). On the other hand, tropical seas are still protected. The greatest intensity of biological activity is located from 10 to 15 ft (3 to 5 m) down. Indeed, coral reefs, which form an extremely favourable environment for the spread of animal life, need light to grow. Beyond 65–100 ft (20–30 m), only shattered madreporian branches are still extant. A dive in tropical seas should therefore be made at around 15–30 ft (5–10 m) down. At such depths, it is possible to stay underwater for two hours with an air cylinder, without having to undergo any stages of decompression. Nevertheless, a decompressimeter is useful when successive dives are to be made.

The Mediterranean is a temperate sea, which is cold, and sometimes extremely cold, for three or four months in the year. Warm clothing is a necessity: 3/16 in (4–5 mm) thick, lined with very flexible nylon foam. The Mediterranean is an almost enclosed sea without tides. There are no heavy currents, except around some little islands like the Planier or the Cassidaigne lighthouses off Marseilles. In these cases, if stages of decompression are necessary, the diver must either keep a close eye on the anchor of the boat or else have much experience in swimming with a compass.

The objectives of diving can be either archeological or directed towards fauna and flora. In the Mediterranean, the diver must descend to between 100 and 165 ft (30 and 50 m) both to reach shipwrecks and to take fishes by surprise. The choice of a diving site is of prime importance. For the most part, the thrill of the dive is directly related to the dangers encountered. In fact, the greater the current and the further the diver ventures towards the open sea, the more rich and interesting becomes the fauna. In tropical seas, it is generally in the channels leading to the coral reefs and along the barrier reefs on the edge of the open sea, that the most enthralling dives are to be made. An extraordinary variety of lagoon and pelagic fauna lives in the channels. However, this should not make the diver forget the dangers, and above all the risks entailed by the current which rises at the beginning and end of each tide, reaching a speed of several knots. It is therefore essential to make enquiries on the whereabouts of these currents and to break off a dive as soon as the slightest difficulty in struggling with them is experienced. Dives along the barrier reefs, close to the open sea, the richer and more interesting and the current is usually less strong. But it is important to beware of the ocean swell, which often makes the approaches to the reef dangerous. Even though, for negotiating a channel, a canoe is sufficient for sheltering behind a reef; for diving at the edge of the open sea, a good boat, and if possible, two crew men are recommended. If one is not thoroughly acquainted with the diving conditions of the locality, one should absolutely avoid having to make stages of decompression: 5 to 10 minutes per stage in a current of 3 or 4 knots dangerously imperils a diver's return to the boat riding at anchor.

Typical scenery on a tropical reef in the Indian Ocean.
In the foreground a shoal of fusiliers can be seen
surrounded by gorgonians.

The Aquarium

The aquarium enables the diver, the fisherman, or simply those interested in marine life to have an unlimited source of interest and amusement in their own home. It has in addition an educational role to play that is particularly important at a time when direct contact with nature is becoming less and less common. The enthusiast is often afraid that he will encounter many problems in keeping marine creatures. Yet today it is easy to set up a sea water aquarium. The scientific study of the conditions of life in an aquarium is very advanced. The techniques and equipment recommended by the specialists are very varied, and can sometimes appear complex. A simple introduction, which will enable the beginner to successfully set up a marine aquarium, will therefore suffice.

Balancing the environment

The intensity of light, the motion of the water, its chemical composition, its temperature, and its degree of acidity are the main factors affecting the balance of the environment in which the animals (fishes and invertebrates) in the aquarium live. In nature, these only vary slightly. The primary concern of the aquarist is to maintain in his aquarium as many of these factors as possible in their normal state. An hermetically sealed tank full of pure sea water will retain its initial characteristics indefinitely. The introduction of a living organism will quickly and drastically alter this balance. In order to exist, an animal must constantly draw some substances from its environment and return others to it. In the water it finds oxygen to enable it to breathe, mineral salts which are essential to growth, and living animals on which it feeds. It excretes solid waste (undigested remains), waste gases (carbon dioxide given off by respiration), and soluble ammoniacal salts. In this way, the gradual disappearance of consumed substances and, conversely, the accummula-

tion of waste products, can be witnessed. The equilibrium is upset and, if the causes of this imbalance are not contended with, the environment will quickly become unsuitable for life. The preservation of this natural equilibrium is extremely complicated and its mechanism is still imperfectly understood. Living organisms which inhabit the ocean are classified into three groups: the producers, the consumers and the reducers.

The producers
These are the plants i.e. algae. The most common are those macroscopic algae which live attached to rocks, but there are vast numbers of microscopic algae inside plankton. These plants absorb water, carbon dioxide and mineral salts, with which, through their ability to photosynthesise, they build up organic molecules (sugars, fats and proteins) and give off oxygen as a by-product. The energy required for this process is derived from sunlight through the agency of chlorophyll.

The consumers
These are the animals. They can only feed on organic compounds (living matter). The herbivores, which are in the majority, are prey for the carnivores. The herbivore group mainly comprises invertebrates (worms, crustaceans, molluscs); fishes are mostly carnivorous. However, in warm seas, there are a large number of herbivorous fishes (e.g. surgeon-fishes).

The reducers
These are the bacteria. Myriads of bacteria quickly attack dead organisms and organic waste products, decomposing them and converting them into mineral salts. Producers, consumers and reducers form a nutritional chain whose self-regulating mechanisms enable a biological equilibrium to be maintained.

A shoal of sweet-lips symbolises the intelligent life of an immensely rich underwater world.

Reconstruction of the marine environment

There are three methods of creating a marine environment in an aquarium. The 'natural' method tries to emulate the mechanisms which, in nature, normally preserve the balance of the environment. This method requires much experience, especially in choosing the inhabitants; so the beginner is advised against it.

The 'semi-natural' method is mainly distinguished from the previous one by the addition of a filtration system which is made from a layer of sand and gravel covering the bottom of the aquarium. This type of filter possesses such a great capacity to nitrify that it is practically impossible for toxic products to accumulate in the water.

The 'artificial' method consists of using certain apparatus to recreate a marine environment and to keep it in equilibrium. Apart from a thermostatically controlled heater and an air pump used in the above methods, there is usually an outside charcoal filter, as well as a powerful pump to circulate the water, a skimmer to clear the decomposed organic matter, an ozoniser, and an ultra-violet lamp to sterilise the water. This method is expensive and makes it difficult to keep some marine animals.

Setting up and maintenance

The cemented plate glass technique is easy, quick, particularly pleasing to the eye, and, above all, it is not too much trouble. A 30 gallon (150 litre) tank can be easily constructed. The different components should be carefully cleaned, and all traces of grease removed with carbon tetrachloride, whilst avoiding all contact with the hands. A silicone elastomer, which does not need special preparation, can be used to assemble the tank. After it has been tested for water-tightness, the cemented plate glass tank can be installed on a completely level stand. A $\frac{3}{4}$ in (2 cm) thick layer of plastic foam will allow the pressure to be spread more evenly. The tank should face the east in order to receive one or two hours of sunlight per day, thus avoiding excessive heat and the growth of too many algae.

Aeration and filtration are of prime importance to the well-being of the prospective occupants of the aquarium. The filter is composed of a bed of gravel or small, carefully washed pebbles and a layer, $1\frac{1}{2}$–2 in (4–5 cm) thick, of well-compressed sand. The sand should be fairly coarse and contain a high percentage of limestone, in the form of crushed shells or coralline sand, for example. This layer covers the strainer which drains the water to the base of the airlift pump, made from a vertical tube through which air is piped down to ensure constant aeration of the sediment. The air is provided by two self-lubricating diaphragm pumps. This type of 'sub-sand' filter is the simplest and the most effective. It can function for years without any attention. By its mechanical action, the filter retains the large impurities, so ensuring that the water is perfectly clear: this is essential for the well-being of the fishes. The innumerable organisms which live in the layer of sand and gravel belong to the reducer group, and feed on waste products. Permanent oxygenation, which favours the growth of certain bacteria, prevents the accummulation of toxic waste. As the surface area of the filter is large, the suction force is relatively feeble, and therefore microscopic animals tend to remain in the plankton. It is essential that the filter is in continuous operation.

Once the filtration and aeration systems are installed, the aquarium may be filled either with sea water, taken directly from an uncontaminated area, or with artificial sea water bought from a shop. The latter, which is freshly prepared, is relatively acidic for about 10 days. The average salinity of sea water is

Biological cycle

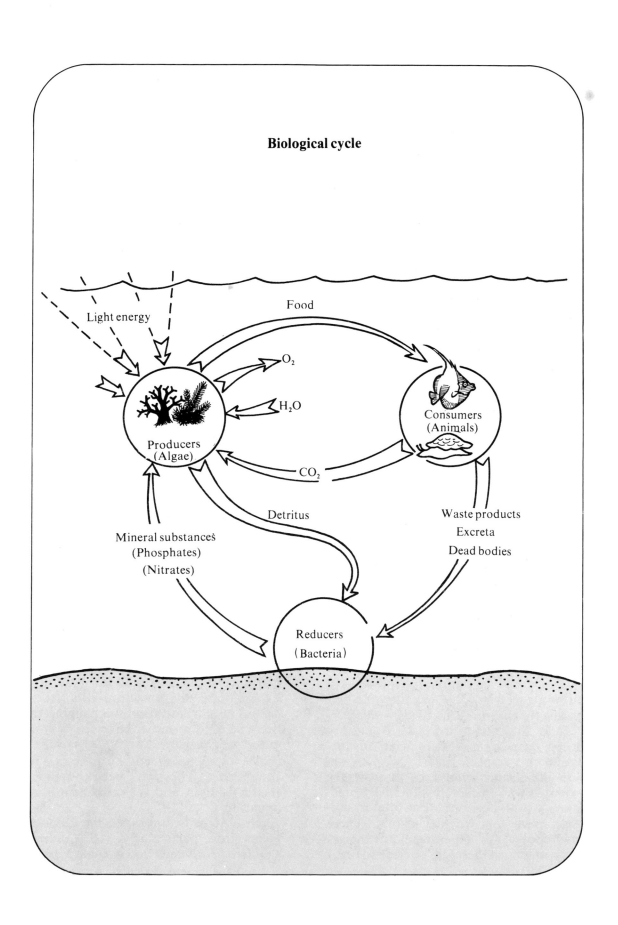

Killer whales: packs of these terrify-
ing marine carnivores attack pods
of Caleen whales.

index

214